Dilemmas in Responsible Investment

DILEMMAS
IN RESPONSIBLE
INVESTMENT

Céline Louche and Steve Lydenberg

Greenleaf
PUBLISHING

Published by Greenleaf Publishing Limited
Aizlewood's Mill
Nursery Street
Sheffield S3 8GG
UK
www.greenleaf-publishing.com

Printed in Great Britain on acid-free paper by
Antony Rowe Ltd, Chippenham and Eastbourne

FSC
www.fsc.org
MIX
Paper from
responsible sources
FSC® C013604

Cover by LaliAbril.com

British Library Cataloguing in Publication Data:

Louche, Celine.
    Dilemmas in responsible investment. -- (Responsible
    investment ; v. 2)
    1. Investments--Decision making. 2. Investments--Moral
    and ethical aspects.
    I. Title II. Series III. Lydenberg, Steven D.
    174.9'3326-dc22

ISBN-13: 9781906093518

Contents

Acronyms and abbreviations

ADEME	Agence de l'Environnement et de la Maîtrise d'Energie
BBBEE	Broad-Based Black Economic Empowerment Act
CalPERS	California Public Employees Retirement System
CDP	Carbon Disclosure Project
CSR	corporate social responsibility
ESG	environmental, societal and governance
Eurosif	European Sustainable Investment Forum
GMO	genetically modified organism
GRI	Global Reporting Initiative
ICCR	Interfaith Center on Corporate Responsibility
IMF	International Monetary Fund
JSE	Johannesburg Stock Exchange
OPERS	Ohio Public Employees Retirement System
ORSE	Observatoire sur la Responsabilité Sociétale des Entreprises
PRI	Principles for Responsible Investment
RAN	Rainforest Action Network
RI	responsible investment
US SIF	United States Social Investment Forum
UKSIF	United Kingdom Sustainable Investment and Finance (previously UK Social Investment Forum)
VBDO	Dutch Association of Investors for Sustainable Development

Preface

It originally seemed like a simple task. In the spring of 2007, we set out to prepare a short 'quiz' on responsible investment. Céline was teaching a course on corporate social responsibility that autumn. For a week-long segment on responsible investment she wanted to identify a number of the basic, but difficult, questions that arise for practitioners in the field.

We weren't certain what we meant by a 'quiz', but we started to jot down some questions from our personal experience. Quickly the questions expanded into short hypothetical stories to dramatise the 'dilemmas', as we were now calling them, that responsible investment practitioners face as they research societal and environmental records of publicly traded corporations, interact with their clients and make investment decisions.

Ultimately, we turned 12 of these dilemmas into the short narratives you find in this book. These 'mini-cases', drawn from our personal knowledge of, and experiences in, the field, were not full-blown case studies of the type frequently used in business schools, but we felt they were detailed enough to promote classroom discussion. In fact, they did just that when Céline taught a number of them during the 2007–08 academic year. But her students raised two interesting questions. The first was, 'What are some real-life examples of these dilemmas?' The other was, 'How would investment professionals in the responsible investment field respond?'

These seemed like reasonable questions and, after some discussion, we decided to develop the cases further by interviewing a number of professionals to find out how they would in fact respond, to track down real-life examples to illustrate the cases, and to provide a general discussion of the various approaches responsible investors take when confronted with these dilemmas, factors that affect these approaches and recommendations about how to proceed when these dilemmas occur.

We now had something that looked more like a book than a quiz. In certain senses it provided a general overview of the field and our views of what makes responsible investment an important and valuable contribution to the financial world. While elaborating on these cases and the field of responsible investment in general, however, we wanted to be careful not to 'explain away' the dilemmas or suggest that they had solutions. We focused on these dilemmas to begin with because we believe that they not only dramatise the complexities of responsible investment, but also demonstrate how contending with these complexities has made valuable contributions to its advancement.

These are not dilemmas that can be resolved, nor should they be. It is in fact in the grappling with the often difficult questions about the relationship between business and society, between the process of investment and its societal and environmental implications, that responsible investment adds its greatest value. It is these hard questions, which do not have easy answers, that lie at the heart of what we do.

Acknowledgements

This book could not have been written without the generosity, support and help of a great many people. Although our two names appear on the front page, many others should be acknowledged for their direct, indirect, always valuable and sometimes challenging contributions.

We are extremely grateful to the 35 people who agreed to be interviewed and to share their insights and knowledge with us. Their thoughts and reactions helped deepen our own understanding of these cases – and we hope will deepen that of the readers of this book as well. Their support went beyond their simple input to the cases. Their enthusiasm and cooperation also provided us with the impetus and energy to complete the book: Eiichiro Adachi, Stewart Armer, Robert Barrington, Mark Bateman, Hans-Ulrich Beck, Seb Beloe, Erol Bilecen, Mark Bytheway, Mark Campanale, Stu Dalheim, Jean-Philippe Desmartin, Gregory Elders, Magnus Furugard, Shin Furuya, Julie Gorte, Gaëtan Herinckx, Geert Heuninck, Stephen Hine, Emma Howard Boyd, Harry Hummels, Jennifer Kozak, Lloyd Kurtz, Tomohiro Matsuoka, Mary Jane McQuillen, Bill Page, Guo Peiyuan, Eckhard Plinke, Kevin Ranney, Mark Regier, Karen Ri, Paolo Sardi, Heidi Soumerai, Raj Thamotheram, Rosl Veltmeijer and Laura Würtenberger.

A number of people helped in the research phase of the project from collecting data to interviewing people or testing the cases. We would like to thank Atlee McFellen who helped track down details on the real-life cases, Katie Grace who continued work on the real-life cases, summarising them for this book, Melissa Tritter who conducted many of the interviews, and Professor Karl Einolf from Mount Saint Mary's University in the

United States and Hager Jemel from Sup de Co Amiens Picardie in France who both tested some of cases with their students.

We would like to express our gratitude to the people who have kindly provided us with comments and reviews. This includes Jette Steen Knudsen, Christopher Voisey, Linda Markowitz and Tessa Hebb.

We are tremendously appreciative of our publisher Greenleaf, who has strongly supported our project from the start, while it was still an idea. We would like to thank especially John Stuart and Rory Sullivan for their patience, help and encouraging words and for their insightful comments, which have been decisive in shaping the book.

The first chapter of this book, 'An overview of responsible investment', first appeared in a somewhat different form in *Finance Ethics: Critical Issues in Theory and Practice* (John Wiley & Sons, 2010). Our thanks to John Wiley and the book's editor, John R. Boatright, for permission to reproduce it here.

Finally, we would like to thank Vlerick Leuven Gent Management School, Domini Social Investments and the Initiative for Responsible Investment at the Harvard Kennedy School for their support of the authors as they undertook this project.

For all these varied and helpful contributions, we are grateful.

Introduction

- This book examines a number of the dilemmas and practical problems responsible investment practitioners face daily when implementing their decisions.

- Our approach emphasises the importance of questions as well as answers and of process as well as product. We stress diversity of opinion and variety of approach.

- We raise fundamental questions about the purpose of investment and the responsibilities of investors, both economic and societal.

Imagine you are an RI money manager . . .

. . . One of your clients is asking you to sell her holdings in a company because it has been accused in the press of contracting with suppliers that have abusive labour conditions . . .

. . . You have to evaluate and benchmark the CSR performance of a number of companies from the same industry, but among them there are companies, primarily the smallest, that provide little or no CSR information . . .

. . . One of your major clients is asking you to exclude companies involved in nanotechnology . . .

. . . what would you do?

Responsible investment – the integration of environmental, societal and governance (ESG) issues into investment decision-making – can be a difficult and complex task. Including or excluding companies, engaging with companies, partnering with stakeholders, evaluating environmental and societal controversies, defining criteria and, all the while, producing a competitive return for investors . . . these tasks can raise multiple questions that cannot be dealt with simply. The practice of responsible investment inevitably raises many such dilemmas as it seeks to balance the competing goals of business, society and finance and to judge how best to reconcile what are often conflicting concerns.

Although these dilemmas are not always easily resolved, we believe that they are also a source of valuable and necessary debate about the appropriate role of corporations in society and the ability of the financial markets to serve appropriately the society within which they operate. It is important that investors acknowledge these dilemmas and participate through a variety of means in these debates. Although they may not have simple solutions, these dilemmas provide a valuable framework for public debate and can encourage the emergence of innovative answers and approaches. Responsible investors join in these debates when they:

- Examine the societal and environmental implications of business activities, actions and behaviour

- Facilitate dialogue between corporations and their stakeholders

- Encourage corporate transparency on societal and environmental issues
- Reward companies that are making genuine efforts towards sustainability
- Integrate societal and environmental data into financial analysis

It is our belief that the acknowledgement and confrontation of these dilemmas position responsible investors at the heart of many of these debates and allow them to participate in important innovations in business and finance. Contending with these dilemmas helps enrich the investment profession and ultimately contributes long-term rewards to society.

Responsible investment: the broader picture

Responsible investment (RI) in its modern form emerged in the 1970s when societal and environmental activists in the United States – concerned with the war in Vietnam, the out-of-control nuclear arms race, the civil rights of African Americans, Native Americans, Hispanics, women, gays and lesbians, and others who had been systematically discriminated against, and the dramatic environmental degradation of the air, water and land – combined the divestment techniques of faith-based organisations that refused as a matter of principle to profit from businesses they regarded as unethical (primarily tobacco, alcohol and gambling) with the tactics of community and consumer activists such as Saul Alinsky and Ralph Nader, who took protests against corporate America from the streets onto the floors of companies' annual meetings. RI's purpose was to communicate to an often uncomprehending and intransigent world of corporate managers the cares and concerns of society when it came to matters of social justice and the environment.

This introduction of societal and environmental concerns into a world of finance that prided itself on its single-minded focus on profits and financial returns was generally ignored by Wall Street or, when it was recognised at all, was greeted with puzzlement at best and, more frequently, with scepticism or derision.

Forty years later, responsible investment enjoys a measure of respectability as it has evolved into a sophisticated, multifaceted discipline. Many of the largest pension funds and institutional investors around the world have now taken up its strategies and tactics. Along with its sister discipline – corporate social responsibility (CSR) – it has found a seat at the table of the mainstream financial community.

As responsible investors attempt to cope with an ever-proliferating range of concerns about business and society and an ever-evolving variety of approaches to incorporating these concerns into investment practices, a number of dilemmas arise. Through a series of 12 short case studies, this book examines these challenges, raising questions about the relationship between business and society, about the purpose of investment and about the responsibilities of investors to various segments of society and the environment. These dilemmas are, to a certain extent, general and abstract, but in practice they take particular forms. The case studies discussed in this book provide specific examples of dilemmas and use these dilemmas in turn to introduce broader issues.

The practitioners' voice

The 12 cases presented in this book are supplemented with a variety of quotations and hypothetical responses from practitioners in the investment community. These quotations and responses are drawn from a series of interviews that we conducted with 35 responsible investment fund managers, researchers and analysts between March and June 2008. Interviewees were from 14 different countries: 12 from Europe, seven from the United Kingdom, nine from North America and seven from Asia.

Each of the 35 analysts and fund managers interviewed were asked to respond to three to four cases. Each case consequently was discussed on average 11 times. Interviews were primarily conducted by telephone and all interviews were recorded and transcribed.

These interviews are the basis for both short quotations (to be found in the 'Responses from practitioners' part of each of the chapters) and for longer hypothetical memos, letters and notes that we imagine might have been written by these practitioners. The latter are based on the responses from these interviews, but we have selectively excerpted, modified and altered the original responses to highlight points we felt were particularly crucial and to provide a cleaner narrative and more coherent story. These texts as they stand here are our versions of what those in the field *might* say – in effect semi-fictional accounts – and are presented here to dramatise key points.

In all cases, those interviewed were speaking anonymously and personally, not as representatives of their firms. We are enormously grateful to these members of the RI community for their willingness to share their thoughts on some of the more difficult challenges we all face and to take time from their busy schedules to support this project. We hope that in the aggregate we have fairly represented their collective wisdom and expertise here.

What can be found in the cases

Each of the 12 cases in this book focuses on a specific key RI theme (see Table 1). These cases often encompass multiple dilemmas and, although presented independently from one another, frequently raise interrelated concerns.

Table 1 **Key themes developed in the cases**

Case	Title	Theme	Related cases
1	Types of responsible investor	Need for customisation	8
2	Ethics and facts	Importance of ethics	7; 9
3	Influence through voice and exit	Means of influence	6; 7; 9; 10
4	Societal returns versus financial returns	Role of the long term	10
5	Alleged versus confirmed illegal activity	Importance of judgement	6; 11; 12
6	When a company changes	Problems with evaluation of change	3; 5; 9
7	Public versus private partnerships for engagement	Building trust	2; 3
8	Relativity of responsible investment standards	Tendency towards the local	1
9	Incomplete societal and environmental data	Challenges in decision-making	2; 3; 6; 11
10	Exclusion of industries	Justification of standards	3; 4; 5; 12
11	Emerging issues	Inevitability of speculation	5; 9
12	Privatisation of public services	Role of government	5; 10

Major topics and themes

The 12 cases in this book touch on the following major topics and themes.

Case 1: Types of responsible investor

Responsible investors do not represent a single and homogenous group. Just like all investors, they have differing risk tolerances and financial needs. But they also bring to the discipline a tremendous variety of societal and environmental concerns that need to be reflected in their investment policies and practices. This case deals with the diversity and variety of responsible investors and the related challenges of product customisation.

Case 2: Ethics and facts

Responsible investment confronts issues that are often highly controversial, such as human rights, child labour and environmental degradation. These are issues about which clients are often passionate. How should the emotional responses to these issues that often arise be dealt with? By what standards or measurements should decisions be made on ethical issues that do not lend themselves to the cold calculus of price and financial returns? This case deals with the challenge of incorporating questions of ethical perceptions and their related emotional responses into the investment process.

Case 3: Influence through voice and exit

What should responsible investors do when dialogue and communications with companies on societal and environmental issues do not lead to satisfactory change? Are there alternatives to engagement in such situations? Should responsible investors sell their stock rather than continue unproductive dialogue? This case explores the issue of influence through exit and voice. Both approaches have advantages and disadvantages – a potential for impact, but with limitations.

Case 4: Societal returns versus financial returns

Responsible investors are entitled to understand what societal or environmental returns they are receiving, in addition to their financial returns. But how can societal and environmental returns be defined, measured and reported? How can the societal and environmental be weighed against the financial? Can trade-offs be made? This case addresses the tension between societal and financial returns, the challenges of measuring and quantifying non-financial issues and the difficulty of balancing short-term and long-term perspectives.

Case 5: Alleged versus confirmed illegal activity

Corporations are frequently accused of breaking the law, sometimes fairly and sometimes unfairly. Courts can take a long time to rule on these accusations, and often the parties settle without admissions of guilt. How should responsible investors respond in cases of alleged illegal activity? Should they wait before acting? Should they make their judgements without a court decision? This case raises the issue of how best to exercise judgement in such situations. It highlights the difficulties of relying solely on the law to judge the appropriateness of corporate behaviour.

Case 6: When a company changes

A company that has had a history of questionable societal and environmental behaviour may change and become an exemplary actor. If responsible investors believe this is possible, they must then assess how significant, far-reaching and long-lasting changes are at firms as they unfold. Some apparent changes wither and die before being fully realised. Other changes unexpectedly become the norm. This case looks at the dilemma faced by responsible investors in determining the meaningfulness of change and deciding at what point it is appropriate to take action.

Case 7: Public versus private partnerships for engagement

Responsible investment can require a sophisticated analysis of how to communicate most effectively with companies to encourage change. But what should responsible investors do when a 'behind closed doors' dialogue does not deliver the desired results? Should they join with others to criticise the firm? Are there other forms of collective action that can help speed change? This case investigates the role of partnerships and public versus private communications as responsible investors seek the best means to effect change in corporate policies and practices.

Case 8: Relativity of responsible investment standards

A tension often exists between the local and the global in responsible investment. Can one apply the same RI criteria and principles regardless of region or do they need to be adapted to local norms? What are the implications for global financial institutions of the tailoring of RI criteria to local standards? This case raises the question of whether responsible investment standards can be viewed as culturally relative or whether they should be treated as universally applicable – or, similarly, whether a single responsible investor can legitimately apply different standards and criteria to a range of product offerings.

Case 9: Incomplete societal and environmental data

It is difficult to practise RI without societal and environmental data. Although the quantity and quality of such information has increased dramatically in recent years, many companies still do not disclose ESG data. This is especially true for smaller companies. This case explores the dilemma of how best to assess companies in the face of incomplete or inconsistent information and the question of whether companies' size or the nature of their business should be factored into the evaluation.

Case 10: Exclusion of industries

One common RI practice is to exclude all companies in certain sectors such as the weapons, tobacco or gambling industries. Similarly, substantial numbers of companies can be excluded because of controversies such as human rights abuses, environmental violations or product safety problems. Is excluding all, or a large part, of an industry an appropriate investment practice? This case looks at issues relating to the justification, definition and implications of excluding industries or companies on the basis of societal and environmental criteria.

Case 11: Emerging issues

Many issues considered by responsible investors involve scientific uncertainties, particularly emerging technologies. Responsible investors can push for additional facts to reduce these uncertainties, but they also recognise that a certain amount of speculation about the future is inevitable. How can one make decisions in the face of these sometimes irresolvable uncertainties? This case addresses dilemmas that arise in assessing the societal and environmental risks and opportunities of new technologies.

Case 12: Privatisation of public services

Over the past 30 years, many national governments have chosen to deregulate or privatise various industries including the telecommunications, financial services, transportation, energy and numerous other sectors. Debates have raged over which services should or should not be privatised or deregulated and to what degree. How should responsible investors evaluate or consider privatisation? Is it part of responsible investors' role to enter into political debates and consider the appropriate relative roles of governments and private enterprise? This case addresses the question of when and how responsible investors should enter into debates that are essentially political by nature.

Table 2 **Key stakeholders and key aspects**

	Case											
	1	2	3	4	5	6	7	8	9	10	11	12
Key stakeholders												
Investors/clients	✓	✓		✓				✓				
NGOs		✓					✓					
Media		✓										
Companies	✓	✓	✓	✓	✓	✓			✓	✓	✓	✓
Industries			✓							✓		✓
Government				✓								✓
Key aspects												
Engagement		✓					✓					
Criteria										✓	✓	
Product design	✓							✓				
Ethics		✓						✓				✓
Transparency		✓			✓				✓			
Measurement and assessment		✓			✓	✓			✓	✓	✓	✓
Partnership							✓					
Financial impact					✓					✓	✓	

Key stakeholders and key aspects

The 12 cases vary as to which key RI stakeholders and fundamental aspects of the RI process they focus on. Most of the cases, except for Cases 1 and 8, focus on multiple stakeholders. However, the stakeholders emphasised in each case differ. Corporations are a focus of all cases except Cases 1 and 8, which focus essentially on investors. Governments are addressed most directly in Cases 5 and 12, the media and NGOs in Cases 2 and 7, and industries in Cases 4, 10 and 12.

With regard to the key aspects of RI, multiple aspects are addressed in most cases. The question of RI measurements is addressed particularly frequently (Cases 2, 4, 5, 6, 9, 10, 11 and 13). Engagement is emphasised especially in Cases 3 and 7. RI standards and criteria, although indirectly addressed in most cases, are explicitly discussed in Cases 10 and 11. RI product design is the focus of Cases 1 and 8. Ethics and its relationship to RI are explicitly addressed in Cases 2, 8 and 12. Transparency is most prominently featured in Cases 2, 5 and 9. The question of partnerships is a key aspect of Case 7.

Table 2 provides an overview of which key RI stakeholders and key aspects of RI are addressed in which cases.

How to approach this book

After this introductory chapter, this book is divided into 14 additional chapters. The first chapter provides an overview of the responsible investment world, its history, primary strategies and techniques, and major players. The following 12 chapters present and discuss the 12 cases with their RI dilemmas. These chapters are organised as described below. Finally, the last chapter provides concluding comments.

The 12 short cases in this book can be read independently and in any order. However, we strongly recommend first reading the Overview chapter, which provides background to the responsible investment field. This chapter provides general information on responsible investment, its history and development, explanations of key terms used throughout the cases (for example, engagement, exclusionary criteria and best in class) and a guide to the different actors involved in the field. This background is important for a full appreciation of the issues and dilemmas in their various forms as raised in the 12 cases.

The cases can be used for different purposes including teaching, professional training or general discussions. We have found them particularly useful in the classroom setting, where they can generate lively debate.

Each of the 12 case-study chapters in this book follows a similar structure consisting of nine separate parts.

The case

Each chapter starts with the statement of a hypothetical case. These cases describe briefly the nature of a particular dilemma and the reader is asked to put himself or herself in the position of a responsible investment professional confronting this question. Although brief, these cases can raise multiple related issues. We have deliberately kept them free of extensive detail to encourage the reader to explore the various possible variations on the basic themes they raise.

Dilemma for the responsible investor

Each case is followed by a section in which we draw attention to the crux of the dilemmas raised and compare the dilemmas briefly to similar issues as they arise in the mainstream financial community. We then highlight some of the implications and nuances of the dilemmas as we see them. These highlights steer the reader to issues that can serve as the basis for further analysis and discussion.

In each of the cases, in this section we have inserted a sidebar called 'The case in perspective'. In this sidebar we summarise and put into perspective the essence of the dilemmas. We also point out why the dilemmas are not easily resolved and how confronting these dilemmas in daily practice has helped responsible investors advance their discipline.

Approaches available to the responsible investor

Although there is no single 'right' solution to the dilemmas raised by these cases, we explore in this next section some of the varied approaches open to responsible investors. Here we sketch the outline of paths that might be taken or scenarios that might be followed as responsible investors confront the dilemma. The potential advantages and disadvantages, and strengths and limitations, of these approaches are set out here for the reader's consideration.

Variable factors

Because these dilemmas have been presented in generalised form, different solutions would be appropriate if different factors should come into play. This section calls attention to some of these different factors and helps the reader explore the types of response that might result under different circumstances. The factors elaborated here are suggestive, rather than comprehensive, and the reader will want to consider additional factors that might potentially impact the response to the case.

Recommendations

This section offers a list of key recommendations that might be followed by those responding to the dilemma raised in the case. They highlight key actions that should be taken.

Graphic

After the Recommendations we provide a graphical representation of the elements discussed in the case. These figures summarise key aspects of the case, sometimes emphasising one or two specific points raised or guiding the reader through the most important questions raised by the case.

Responses from practitioners

As well as some quotations from responsible investors, this section features our hypothetical responses of practitioners. The responses here have been drawn from the interviews we conducted but shaped to dramatise certain points and positions. For each case, we selected two or three responses provided by our interviewees and, working from their responses, composed fictitious minutes, notes, memoranda and letters. The 'minutes' are summaries of imaginary meetings at a financial services firm. The 'notes' represent an analyst's or manager's thoughts set down in preparation for a meeting. The hypothetical inter-office 'memos' imagine how one professional might address the issue with another at his or her firm. Finally, the 'letters' imagine correspondence that might take place between a money manager and a client to explain various positions. In this section, the reader may want to compare and contrast the different responses from practitioners with his or her own responses.

In the news

This section provides details of a well-publicised event that exemplifies the type of dilemma described in the case. This brief account is intended to draw the reader's attention to one particular manifestation of the specific kinds of complication that can arise.

Cases for comparison

Many of the cases presented here are interrelated, addressing different aspects of similar issues. In this final section we note cases where the reader might benefit from comparisons between cases. These comparisons can help the reader explore related points between contrasting cases.

Conclusion

Our hope is that this series of cases can promote the kind of dialogue, debate, collaborative learning and collective thought that will advance the theory and practice of responsible investment as an emerging discipline.

For students of business and finance unfamiliar with the field, these cases can provide an introduction to RI's various practices. For finance professionals entering the field, they can provide an opportunity to explore some of the challenges they will face daily. For academics, they can provide a useful guide to the complexities of responsible investment and its similarities and differences with mainstream finance. For the general public, they can serve as introduction to an emerging theme in investment that receives considerable attention in the press.

The cases in this book are part of the story of responsible investment as it has evolved in the late 20th and early 21st centuries. The dilemmas portrayed in these cases arise in part because responsible investors are breaking new ground – incorporating societal and environmental factors into the investment process, promoting disclosure of corporate social responsibility data and seeking to direct corporate practices towards the creation of a more sustainable and just world.

As the field continues to grow, responsible investment will develop and refine its practices in these areas, although the dilemmas themselves will not disappear. It is in grappling daily with dilemmas such as these that responsible investment will continue to develop and enhance the abilities of finance to bring long-term benefits to society and the environment.

An overview of responsible investment

- **What is responsible investment and where does it come from?**
- **What are the practices of responsible investment?**
- **Who are the major players in the field of responsible investment?**

Responsible investment (RI) is a product, a practice and a process. RI is an investment product in the sense that in addition to financial factors, investors acquire, hold or dispose of companies' shares on the basis of environmental, societal and governance (ESG) factors as well as ethical factors. It is a practice in the sense that RI is a way to identify companies with strong sustainability records and to engage with companies to encourage improved ESG performance. And RI is a process through which it tries to influence corporations' behaviour on a range of societal and environmental issues to move towards more sustainable development.

RI manifests itself in many ways and, not surprisingly, goes by many names – it is variously referred to as socially responsible investing, ethical investing, sustainable investing, triple-bottom-line investing, green investing, best-of-class investing, ESG investing, impact investing and, most simply and more recently, responsible investing.

These different names reflect in part the different approaches of the groups from which it has evolved over the years: religious organisations stressing the ethical aspects; environmental organisations finding comfort in the language of sustainability; labour, human rights and community organisations stressing the societal aspects; and those most concerned about shareholder rights favouring the theme of governance.

Underlying these differing interests and approaches is a common theme – that of long-term value creation. Whatever the particular vocabulary, responsible investors are inclined to focus on the long term rather than the short term and to measure their rewards in values rather than price. Values in this context refers not only to economic value, but to the broader values of fairness, justice and environmental sustainability to which companies and investors can, and indeed do, have an obligation to contribute.

The RI community acknowledges that any investment inevitably has an impact on society or the environment – and it accepts its responsibility to evaluate that impact and to direct it, as much as reasonably possible, to societally productive ends, while achieving competitive returns. Simultaneously, responsible investors understand how data and discussions about corporations' interface with society and the environment can help them better manage risk and make better-informed investment decisions.[1]

Among the key characteristics of RI are the following:

- Responsible investment encourages a long-term perspective. It does so because (1) ESG and ethical issues cannot always be captured by the market, which tends to be short term in its perspective; (2) RI seeks to establish trust between stakeholders (employees, communities, customers, suppliers) and corporations, a trust that results from ongoing dialogue and persists over time; and (3) RI encourages corporate executives themselves to adopt a long-term perspective in managing their company, one that balances societal and environmental concerns with the drive for profits

- RI adopts a stakeholder perspective. At the core of the conception and practice of RI is the belief that all stakeholders in the corporation matter and that a productive and profitable company will invest in its full range of stakeholders, receive a return from all its stakeholders and consequently be able to provide a long-term return to its investors[2]

- Responsible investment encourages interaction between society and corporations. It does so because the market signals sent by price alone are a blunt tool and are often insufficient for nuanced communication. Communicating through market signals alone cannot address many of the intangible values that are important in a properly functioning society

This focus on the long term, the full range of stakeholders and nuanced communication between investors and those with whom they place their assets, makes for a more complicated conception of investment than that practised today by the mainstream. It is far easier to define and measure success in investment by price-based performance than to assess the success of specific investments in adding value and values, both tangible and intangible, to society.

The 12 cases in this volume illustrate the kinds of hard question this focus compels investors to ask. This introductory chapter provides background on the history of how and why responsible investment has come into being, the tools and strategies it has evolved in order to achieve its particular goals, and the major players and current trends in the field as it stands today. This background context will inform a deeper understanding of why certain dilemmas arise for responsible investors as they face these difficult questions and work toward reasonable answers.

History of responsible investment

The concept and practice of responsible investment have evolved over the past 40 years in what can be classified as five primary periods. Each phase of this evolution was characterised by its own particular practices and concerns. Each new phase of RI tended to incorporate previous forms that persisted and coexisted with the new ones.

Phase One: Roots. The earliest stage of responsible investment, before it was known as responsible investment, dates back to the eighteenth century.[3] For several hundred years religious institutions – such as the Society of Friends (Quakers) and the Methodists – were precursors to the modern form of RI in that they believed that investing was not a neutral activity but implied values. They shunned 'sinful' companies whose products conflicted with their basic beliefs. These so-called 'sin stocks' were for the most part those of companies involved in alcohol, tobacco, gambling and, in certain cases, weapons. One of the emblematic and first RI mutual funds of this first phase was the Pioneer Fund, launched in the United States in 1928.

Phase Two: Development. The second phase dates from approximately 1970 and runs through to the late 1980s. It marks the beginnings of RI in the contemporary sense of the term and is typified in the United States by the Pax World Fund, launched in 1971, and in Europe by the Friends Provident Stewardship Unit Trust in 1984.

In the United States, this incarnation of RI originated in part in the political and protest movements of the day. The Vietnam War and apartheid in South Africa were two issues in particular that drove the RI movement of that time. Other citizen movements such as civil rights, women's liberation and the environment raised issues of crucial concern to the RI movement and became a part of its lobbying of corporations on issues seen as unethical.[4]

At that time, Ralph Nader and Saul Alinsky, prominent consumer and community activists of the late 1960s, started to use the shareholder right to appear at corporate annual meetings and to file shareholder resolutions to raise societal and environmental issues directly with corporate management. Nader's General Motors campaign leading to the submission of two socially based resolutions on the annual meeting proxy ballot remains a historic moment, as does Alinsky's leading of community groups into the annual meeting of Eastman Kodak. These tactics were soon adopted by the RI movement and became an important second tool for responsible investors.

In the 1980s, RI also took root in Europe. The Friends Provident Stewardship Unit Trust was among the first ethical investment funds in the United Kingdom and a precursor to many similar funds. A number of eco-banks such as Triodos Bank in the Netherlands were also founded during that time.

Simultaneously, a number of RI support organisations were created, such as the Interfaith Center on Corporate Responsibility (ICCR) in 1971, the first two Social Investment Forums in the United States (1980) and the United Kingdom (1983),[5] as well as the first professional RI rating agencies, such as KLD Research & Analytics (United States, 1988) and EIRIS (United Kingdom, 1983).

During this second period, RI developed in a political climate of social protest and unrest and was transformed from a faith-based activity (using ethical principles in the construction of investment portfolios) into an activity promoting a public awareness of the social responsibility of corporations and of investing (the self-conscious phenomenon of RI).[6]

This was the period in which RI was first used to lobby corporations to adopt responsible and ethical practices.[7] It was driven primarily by a combination of individuals, religious organisations and other mission-driven organisations concerned with social justice and environmental issues.

Phase 3: Transition. During the early 1990s, RI began a gradual transition to a less confrontational approach with a strong growth in environmental concerns. The Brundtland Report, which highlighted and defined the concept of sustainability, was published in 1987.[8] The Kyoto Protocol on climate change was ratified in 1997. For RI, this meant the emergence of so-called green funds, especially in Europe, less concerned with avoidance and ethical issues and stressing identification of specific positive sectors or activities linked to the environment, such as renewable energy and clean technologies. In addition, during the mid-1990s, European governments in particular began to promote the concept of corporate social responsibility and responsible investment, lending increasing legitimacy to the movement.

During this period the number of social rating agencies grew significantly; the first RI index, the Domini 400 Social Index, was launched (1990), and corporate social responsibility (CSR) consultancy organisations such as SustainAbility began to thrive. These developments were typical of the new, more systematic and analytical aspects of this third phase with its emerging emphasis on sustainability and cooperation.

Phase 4: Expansion. The beginning of the 21st century heralded a turning point for RI in both its approach and its growth. This fourth period was characterised by the professionalisation of the field and a growing worldwide interest in its practice. RI began to find acceptance in the mainstream investment community, leaving behind its more activist image and becoming a more commercially viable endeavour.[9] This evolution was closely linked to the growing importance of CSR and the increasing accessibility of CSR reports issued by corporations. The Global Reporting Initiative, launched in 2000, played a crucial role in this growing acceptance of CSR reporting by corporations and the increasing thoroughness and sophistication of these reports.

In the early 2000s, institutional investors started to become broadly involved in RI. This growing interest on the part of institutional investors was partially stimulated by governments in Europe at both the European and national levels. In several European countries, legislation and regulations required pension funds to publicly state the degree (if any) to which they took into account ESG considerations in their investment decisions. In the United Kingdom, RI pensions disclosure regulation was enacted in 2000, followed shortly thereafter by several other European countries. In addition, in 2001 the Norway Petroleum Fund adopted several RI policies, initiating a movement among major European pension funds towards the incorporation of RI practices. In particular, the interest of institutional investors in RI explains the substantial growth of the assets under RI management in Europe during the first decade of the 21st century.

With the increasing involvement of institutional investors, a best-in-class approach to stock selection found growing acceptance. This approach stresses broad diversification (no elimination of industries entirely), positive rankings (only the best companies in each industry are included) and quantitative measurements (companies are scored on sustainability indicators). Companies are evaluated on their performance relative to their peers, rather than in absolute terms. By stressing best practices, the approach promotes societal and financial values and at the same time encourages competition among corporations to achieve societal and environmental goals.

Phase 5: Mainstreaming. As the first decade of the 21st century drew to a close, RI stood at a crossroads. Its increasing acceptance by institutional investors was marked by such events as the launch of the Principles for Responsible Investment (PRI) in 2006. By 2010, the PRI had grown into a coalition of more than 800 of the largest institutional investors and asset managers worldwide, representing some $22 trillion under management.[10] In many senses RI appeared to be poised to become a mainstream investment practice applied across various asset classes. A number of the members of the PRI, including the French national pension fund (Le Fonds de réserve pour les retraites – FRR) and the California Public Employees Retirement System (CalPERS), were extending the concepts of RI to asset classes beyond public equities.

Phase five marks the emergence of a broader-based concern with the integration of many of the interests of responsible investment with mainstream investment practices. Integration of this sort focuses on the consideration of ESG data in the valuation of stocks and the construction of portfolios – in short, in the traditional investment decision-making processes.

At the same time, however, it is true that many of the daily investment practices within the mainstream appear to be increasingly moving in the opposite direction, that is, towards short-term time-horizons, the incorporation of high-risk investments and a razor-sharp and exclusive focus on the financial aspects of the investment process.

Growth of assets and questions of financial performance

Although precise figures are difficult to come by, there has been a significant growth of assets under RI management and the number of RI funds over the years, a growth that, it is not an overstatement to say, has been explosive since the late 1990s. In the United States, RI assets under management were at $2.71 trillion in 2007, representing 11% of the $25.1 trillion in total assets under management.[11] RI assets increased by 324% from 1995 to 2007, a faster growth than the broader universe of all investment assets under professional management. In Europe from 2002 to 2010, RI assets under management have been multiplied by almost 15. Eurosif (European Sustainable Investment Forum) in 2010 placed the value of the RI market at €5 trillion as of 31 December 2009.[12] RI has been experiencing an exponential growth especially between 2007 and 2009 with an 87% growth of the market. According to Eurosif's 2010 figures, RI accounts for 43.1% of total European funds under management.

Throughout the various phases of its development, the question of whether responsible investment imposes costs on financial returns has been the object of ongoing debate and extensive academic study. Advocates of responsible investing have argued that societal and environmental criteria can help investors avoid risks unrecognised by traditional stock analysts, help identify high-quality corporate management and highlight companies that are attuned to emerging issues – all of which should help boost performance.[13] Critics have argued that, according to modern theories of portfolio management, any restriction on a universe of potential investments will increase undiversified risks and reduce risk-adjusted returns.[14] Sceptics have pointed out that assets managed under such criteria are insufficient to move stock prices. Advocates have argued that societal and environmental performance can affect a company's overall reputation and that companies with stronger reputations can command higher price-to-earnings ratios in the stock markets and borrow at lower rates in the bond markets.[15]

Although this debate is likely to continue, considerable research indicates that, in general, societal and environmental screening as recently practised does not hurt a fund's financial performance.[16] For example, a review of 31 socially screened mutual funds from 1990 to 1998 found that on average they outperformed their unscreened peers, but not by a statistically significant margin.[17] Similarly, a 2001 academic review of 80 studies on the links between CSR and financial performance found that 58% of the studies observed a positive relationship to performance, 24% found no relationship, 19% found a mixed relationship, and only 5% found a negative relationship.[18]

Some academics have noted that the 'sin stocks' of tobacco, alcohol and gambling companies have historically performed well, often outperforming the broad markets over long periods of time.[19] For example, Jeremy Siegel in his book *The Future for Investors*[20] points out that the tobacco company Philip Morris (now known as Altria) was the single best performing stock of the 20th century. He attributes this outperformance in part to the company's ability to generate substantial cash flows in good times and bad, due to a loyal customer base, and consequently pay an increasingly generous dividend. Outperformance by a number of tobacco, alcohol and gambling companies, however, does not necessarily mean that a well-diversified portfolio without these stocks cannot perform in line with a similarly well-diversified portfolio with them.

One indication that responsible investment can be managed in such a way as not to impose a long-term cost on investors is the record of the MSCI KLD 400 Social Index, the first of the RI indexes originally launched under the name Domini Social Index in 1990. For the 20 years from its inception on 1 May 1990 to 30 April 2010, this index had an annualised return of 9.70%, as compared with the Standard & Poor's 500 Index that returned 8.88% over this period.

As responsible investment finds application in asset classes beyond publicly traded stocks – such as fixed income, real estate, venture capital and private equity[21] – this debate about the performance of responsible investment in practice will almost certainly continue.[22]

Motivations for responsible investment

There are four major factors that motivate responsible investors, in addition to the desire to receive competitive returns. These motivations are not necessarily mutually exclusive and frequently responsible investors are motivated by two or more simultaneously. This mix of motivations reflects that fact that responsible investment can serve multiple purposes. Although these overlapping and complementary motivations can sometimes be confusing to those outside the discipline and can pose challenges for those within it, they reflect the flexibility and multiplicity of purposes that responsible investment can serve.

Avoid profiting from unethical behaviour. As noted above, this is historically the oldest of the motivations for responsible investors, but is still an important factor for many. The underlying principle here is 'do no harm' and it reflects the basic desire not to profit from products or services that are harmful to society. In addition to those who choose not to profit from products such as tobacco and gambling or weapons of mass destruction, many also strongly reject profiting from companies that ignore child labour or human rights or conduct their business with a blatant disregard for the environment.

Encourage corporations to enter positive lines of business or to develop strong stakeholder relations. Conversely, other responsible investors may actively seek out investments in companies engaged in business lines that have a notably positive story. They may be breaking new ground on the environment in areas such as alternative energy, energy efficiency, pollution control or organic farming. They may be focusing on products and services that promote equality of opportunity such as vaccines, mobile telecommunications, microfinance or clean water. Similarly, responsible investors may be motivated by a desire to hold out for praise and emulation companies that have particularly strong stakeholder relations, investing in their employees, their customers, their suppliers and their communities, and in turn being rewarded by these stakeholders through loyalty and high-quality relationships.

Pick stocks that will outperform and avoid stocks that will underperform. Financial performance remains an important driver for RI fund managers, as investment returns cannot be disregarded. RI may be motivated by a desire to identify societal and environmental factors often overlooked or underappreciated by mainstream financial community – factors that may lead to financial risks or provide opportunities for financial rewards. By incorporating relevant information of this type into their stock valuations, they hope to identify stocks that will outperform over the long run.

Seek to change corporate behaviour. Since the 1970s, the desire to improve corporate behaviour has been a primary motivation of responsible investors. The improvements sought can address many concerns widely recognised as societally and environmentally beneficial in areas as diverse as energy and the environment, fair employment, disclosure of societal and environmental data, corporate governance and labour standards and policies. These responsible investors seek to bring pressure for positive change either directly through dialogue with corporations, or indirectly by selling or refusing to purchase a company's stock.

The basic practices of responsible investment

Responsible investors also employ a variety of practices in seeking to achieve their varied goals. These practices can vary depending on four primary factors: (1) strategic approaches taken, (2) tools used, (3) degree of commitment, and (4) organisational approach adopted. Because these four factors vary from responsible investor to responsible investor, they produce a varied set of approaches to implementation of the discipline that parallels the varied motivations that are also in play. As with the case of motivations, this variety of nuance within RI practice can be confusing to outsiders and pose challenges for insiders. The flexibility it provides, however, strengthens the ability of RI to act effectively within the corporate and financial worlds.

Strategic approaches

Five strategies for implementation of RI standards or criteria coexist within the RI community. They correspond approximately to the four motivations listed above that different responsible investors bring to the practice. These strategies can be used independently or, as often happens, in combination.

- **Avoidance.** This approach corresponds roughly to the motivation of avoiding profits from unethical behaviour. The strategy is to shun investments in companies engaged in businesses or practices regarded as unacceptable or generally harmful to society. It can be based on the exclusion of certain sectors or of certain activities

- **Inclusion.** This approach corresponds roughly to the motivation of encouraging corporations to enter positive lines of business or to develop strong stakeholder relations. Investors adopting this strategy invest in companies engaged in business areas or practices they view as exceptionally beneficial to society or stakeholders

- **Relative selection.** This approach corresponds roughly to the motivation of picking stocks that will outperform and avoiding stocks that will under-perform. Investors adopting this strategy select sector leaders on ESG criteria. They invest across all industries and sectors, selecting the best-performing companies in each

- **Engagement.** This approach corresponds roughly to the motivation of seeking to change corporate behaviour. Investors adopting this strategy engage with companies to voice shareholders' concerns on societal and environmental issues

- **Integration.** This approach corresponds roughly to the motivation of integrating ESG issues into traditional financial analysis. Investors adopting this strategy tend to consider ESG aspects as risks. These investors primarily focus on material ESG issues, that is, issues that are most likely to impact the company's financial performance and stock performance

According to recent surveys, as much as 70% of the American RI industry employs some kind of avoidance strategy,[23] and approximately 37% of the European RI industry uses avoidance criteria.[24] In addition, within the Islamic world a form of responsible investment based on the precepts of the Koran combines an avoidance approach that limits companies involved in finance, as well as alcohol and pork, with a concern for the role of finance in promoting distributive justice. Although the avoidance strategy has been criticised by some as limited in impact and scope and as conveying a negative message that fails to encourage companies in excluded industries to improve their CSR commitments,[25] it remains an important strategy in 2010.

Inclusionary approaches have recently gained in popularity throughout the world with the development of specialised funds such as clean tech, alternative energy and water funds. In Europe, assets in these thematic funds increased by 33% between 2008 and 2010.[26] A number of thematic indexes have also been developed, including the S&P Global Alternative Energy, FTSE Environment 100 Opportunity, ECPI Global Climate Change, Janney Global Water and the Credit Suisse Global Warming Indexes.

The first decade of the 21st century has seen an increasing interest in relative-selection approaches, especially in Europe among institutional investors. And since 2007, the integration strategy has also been gaining ground and is becoming one of the dominant strategies in Europe. According to a Eurosif 2010 report,[27] approximately 57% of the European RI industry use this integrated approach. The PRI plays an important role in this trend through its first principle, which encourages the integration strategy.[28]

Strategies within the responsible investment world have demonstrated a tremendous diversity of approaches. Over time, new approaches have developed and will undoubtedly continue to do so. Although the dominant strategies and approaches vary from region to region, they are all, to one degree or another, practised by members of the responsible investment community throughout the world.

Tools

For each of these five general strategies, a variety of specific tactics has been developed within the RI community since the 1970s.

The **avoidance strategy** has led to the development of several different approaches to the setting of negative or exclusionary standards-based criteria relating to products, services, policies or actions. These standard-setting tools include the following:

- Standards relating to products viewed by investors as harmful or unethical. Some of these products, such as tobacco, alcohol and gambling, have been historically viewed negatively by religious organisations. Similarly, a substantial Islamic finance practice has grown up in recent years and applies standards based on the teachings of the Koran. Among these is one forbidding usury, which in effect excludes most companies in the financial sector

- Standards relating to products viewed by particular segments of the responsible investment market as more generally harmful to society. For some there is a more general agreement on their harmfulness, such as landmines, nuclear weapons, ozone-depleting chemicals and disregard for universal human rights. These standards tend to draw on international treaties or principles endorsed by governments but applicable in general to companies

- Standards relating to controversial products and activities for which a considerable variety of opinions exists, such as pesticides, nuclear power, infant formula or animal testing

- Standards relating to companies doing business in countries generally regarded as contravening international human rights standards, such as Sudan and Burma

This range of screening techniques results in a wide range of responsible investment portfolios, some with fewer exclusions and others with more, a variety that results from the desire of many responsible investors to have the holdings in their portfolios tailored to their individual or institutional concerns.

The **inclusion strategy** has led to the development of a set of positive standards tools that identify companies or sectors particularly beneficial to society:

- Standards identifying companies promoting environmental sustainability through the development of energy efficiency, renewable and alternative energy technologies, pollution control and prevention, public transportation, and similar initiatives

- Standards identifying companies promoting economic development and health among the historically underserved, such as vaccines, mobile telephones, microfinance and micro-insurance, clean water, and related initiatives

The **relative-selection strategy** has led to the development of a best-in-class screening tool. Best in class generally employs a substantial number of ESG criteria to score and rank companies. It then selects the best-performing (for example, top 10%) companies in each industry and excludes the rest. The number of ESG criteria used varies greatly. For example, SAM Group, a long-time proponent of best-in-class screening, uses some 130 criteria. Asset4, acquired by Thomson Reuters in 2010, has developed a methodology that employs approximately 250 key performance indicators, which it uses to rate and rank 2,300 companies worldwide. By contrast, the Swiss-based money management firm Pictet has argued that 'less can be more' and finds that a limited number of key performance indicators – even as few as one or two per industry – can be sufficient to identify the best and worst actors in that industry.

Table 3 provides examples of ESG criteria that are widely used in the RI world and often serve as the basis for best-in-class screening. In constructing scores for ratings, a weighting system is often applied to the different criteria in each sector to reflect the varying degree of importance of ESG issues for different industries. For example, environmental issues may be given substantial weight in the chemical sector, while human resource issues may carry a stronger weight in the computer software industry.

Table 3 **Examples of environmental, societal and governance issues**

Environmental (E)	Societal (S)	Governance (G)
• Emissions	• Stakeholder relations	• Board structure
• Environmental policies	• Working conditions	• Independent directors
• Environmental management systems	• Respect for human rights	• Independent leadership
• Toxic chemicals	• Diversity	• Separation of chairperson and CEO
• Genetic engineering	• Workplace health and safety	• Remuneration
• Pollution	• HIV/AIDS	• Shareholder rights
• Water	• Product safety	• Accounting quality
• Energy efficiency	• Treatment of customers	• Audit quality
• Hazardous and solid waste	• Labour relations	• Board skills

The **engagement strategy** has led to the development of a set of tools that promote and facilitate dialogue with corporations. The three primary modes of RI engagement that have emerged are proxy voting, the filing of shareholder resolutions and direct dialogue with corporations.

Voting on the shareholder resolutions that appear on corporate proxy statements for annual general meetings is not only the right of all stockholders but also the fiduciary obligation of institutional shareholders. Most resolutions appearing on proxy statements relate to corporate governance issues such as election of board members, selection of auditors, approval of compensation packages for executives, and related issues. In addition, well over one hundred shareholder resolutions relating to societal and environmental matters also appear on the proxy statements of companies around the world, but particularly in the United States and

Canada where the filing of such resolutions is relatively easy. Adopting RI voting policies and communicating these policies to corporate management is therefore the most common form of engagement and is essentially applicable to all investors.

Filing shareholder resolutions is a more direct engagement with management, particularly useful in the United States and Canada due to its relative ease of implementation. Since the early 1970s in the United States, members of the Interfaith Center on Corporate Responsibility have filed over 100 such resolutions on societal and environmental issues each year. Increasingly, unions are also using this tactic to raise corporate governance concerns. Although few proposals on societal issues receive majority votes and are generally only advisory in nature, these resolutions are an important tool in reaching management and initiating dialogue.[29]

Dialogue with corporate management is the third widely used form of engagement. RI institutional investors often enter into dialogue with managers on societal and environmental issues such as human rights, labour standards, the environment and diversity, as well as corporate governance matters. Increasingly these dialogues are conducted through coalitions of RI investors, such as those participating in the Carbon Disclosure Project (CDP). The CDP is a coalition of institutional investors urging the largest corporations in the world to measure and disclose their carbon emissions.

Between 30% and 40% of the US and European RI industries currently engage in some kind of shareholder activism or dialogue.[30] However, it is important to note that the issues addressed through engagement differ significantly between countries and even between asset managers. While the US RI industry is particularly active in filing shareholder resolutions and public engagement, Europe is notably active in direct private engagement.[31]

The **integration strategy** has been gaining widespread acceptance in recent years, especially in the French market, due to the involvement of large investors such as public and private pension funds and insurance companies.

The integration approach uses the analysis of ESG and other intangibles to inform the investment process at all levels. It differs from the other approaches in that it does not remove from investment consideration any firms or sectors, nor does it apply a best-of-sector approach, but rather it makes use of these ESG factors to determine companies' profitability prospects and stock valuations.

The integration strategy requires money managers to make direct links between ESG issues and stock valuation. ESG data is not used independently of financial analysis but as a means of enhancing financial analysis directly. ESG provides financial analysts with valuable forward-looking information about corporate operations, innovation, intellectual capital and management quality, as well as potential environmental and societal risks and rewards, and helps them make more complete assumptions about the competitive positioning and future financial performance of these firms.

These RI strategies and tactics, like the motivations for responsible investment, are not mutually exclusive. For example, a single responsible investor might avoid certain industries, use best in class for those not avoided and engage with companies both included and avoided. Because of the wide range of strategy and tactics available to responsible investors, the choices in how best to proceed in the integration of societal and environmental concerns into the management of a particular client's portfolio is not always easy.

Degree of commitment

Further complications can arise depending on whether the responsible investment is applied to all or only part of a client's portfolio. Most stand-alone RI mutual funds in the United States, such as those of Calvert, Domini and Pax World, apply their particular RI standards to all the stocks in their equity funds. Large institutional investors, however, may choose to apply standards to only part of their overall allocation to stocks. For example, in the early 2000s, a number of large European pension funds, including the Dutch funds ABP and PGGM, allocated a limited portion of their equities to sustainability funds with environmental criteria to test the effects of applying RI practices.

Still other large institutional investors may apply responsible investment tools to all their assets, but use different tools for different parts of their portfolios. For example, this approach was promoted in the early 2000s by F&C Investments, which operates one of the largest screened ethical funds in the United Kingdom and for the rest of its unscreened assets adopts an engagement approach only.

Organisational approach

Responsible investors – be they pension funds or foundations, banks, private money managers, or mutual funds and unit trusts – can choose to implement their RI strategies using internal resources or by outsourcing them to vendors. Activities that can be outsourced include research, screening and engagement.

With respect to research, responsible investors can create their own in-house research team to evaluate the societal and environmental records of publicly traded companies, or purchase such services from specialised outside vendors. The advantage of an in-house team is that RI investors can focus their research on the specific CSR issues of greatest importance to themselves; or they can perform customised research for others and thereby facilitate the creation of separate portfolios tailored to the concerns of a variety of clients.

Examples of asset managers with in-house RI research capabilities – sometimes referred to as 'green teams' – include F&C Asset Management and Aviva Management in the United Kingdom; Dexia Asset Management in Belgium and France; and Calvert Asset Management, Domini Social Investments and Pax World Mutual Funds in the United States. These teams typically range in size from 5 to 15 researchers and are charged with developing the RI standards for in-house funds, conducting CSR research on specific companies and maintaining lists of approved and excluded companies. These in-house research teams may supplement their own research with research from one or more outside vendors.

Money managers or asset owners may also decide to rely primarily on outside RI research vendors. In this case, the firm typically maintains an internal staff of one to three persons to oversee and apply the research obtained from these vendors. By using outside vendors, the manager or asset owner avoids the internal decision-making and expense of customised research for standard-setting. This approach is often adopted by money managers wishing to serve an RI clientele along with their other conventional clients, but choosing not to make RI a primary focus.

For engagement, similar options for developing in-house expertise or outsourcing of these services exist. An additional consideration for RI investors undertaking engagement is whether to act alone or in coalitions. As RI becomes an increasingly accepted practice, the trend is towards engagement in broad-based coalitions. One of the largest and most successful of these is the Carbon Disclosure Project, which as of 2010 included some 534 institutional investors with combined assets of

Figure 1 **Map of approaches to responsible investment**

Source: Adapted from C. Louche and S. Lydenberg, 'Responsible Investment', in J.R. Boatright (ed.) *Finance Ethics: Critical Issues in Financial Theory and Practice* (Hoboken, NJ: Wiley, 2010): 406

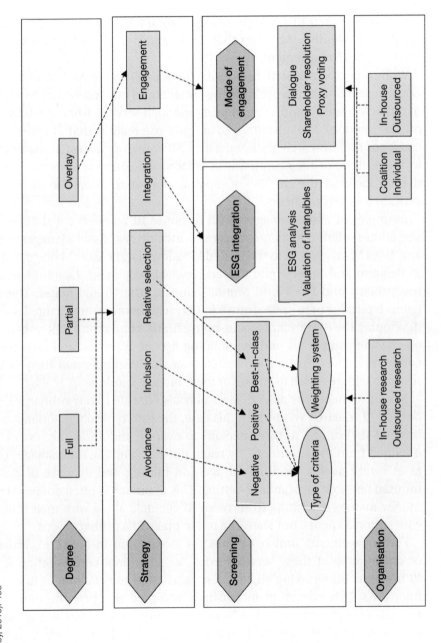

approximately $64 trillion, seeking disclosure from the largest corporations in the world regarding their carbon emissions. In the United States, since the early 1970s, the Interfaith Center on Corporate Responsibility, a coalition of some 300 institutional investors, mainly religious organisations with combined portfolios worth an estimated $45 billion, has served as a coordinating organisation for those engaging with corporations on a wide variety of CSR issues. In the Netherlands, VBDO, the Dutch Association of Investors for Sustainable Development, provides voting services and engages with companies in order to direct corporate policies and behaviour towards sustainable performance. It has provided voting advice to its clients since 2005 and engages in dialogue with publicly traded Dutch companies on their behalf.

Coalitions have the advantages of pooled resources and coordinated action in approaching companies, increasing the chances for a successful engagement. In particular, they raise the level of seriousness of dialogue by assuring corporations of a breadth of concern for specific issues.

The approaches to RI outlined in this section are summarised in Figure 1.

The major players in responsible investment

The diversity of those in the responsible investment movement is both one of its strengths and one of the sources of the challenges its practitioners face daily. The need for this diversity stems in part from the challenges of directing large, publicly traded corporations to act in the public interest. A concerted effort of regulators, non-profit community and environmental organisations, consumers, labour unions and others – combined with the efforts of asset owners, money managers and services providers – is required.

The major players in the world of RI finance are (1) asset owners (individuals, small institutions, and large government and private pension funds adopting RI policies and practices); (2) asset managers (RI specialists, mainstream financial firms with an RI subspecialty, and mutual fund managers, financial advisers and sole practitioners with an RI focus); and (3) providers of support services (RI research firms, RI engagement services and RI associations and think-tanks) (see Fig. 2). Each of these sets of players approaches RI with their own particular perspective and set of needs, creating a complex and varied set of forces that can be brought to bear on corporations in numerous ways.

Figure 2 **Major players in responsible investment**

Asset owners	Asset managers	Providers of support services
Retail investors	Responsible invest-ment specialists	Research firms
Small institutional investors	Mainstream financial firms	Engagement services
Large institutional investors	Financial advisers and consultants	RI associations and think-tanks

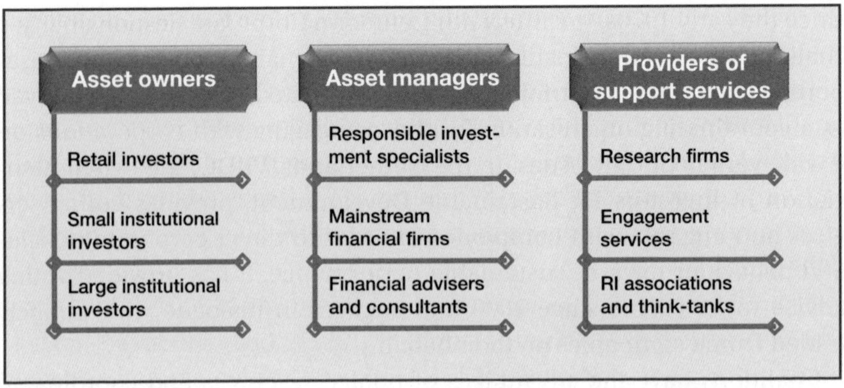

Asset owners

The three types of asset owners that are the primary drivers behind the RI movement – retail investors, small institutions and large government and private pension funds – come to RI with differing perspectives and consequently play differing roles.

Retail investors

Responsible retail investors are individuals wishing to invest in corporations or other assets that have positive societal and environmental impact and to avoid those with questionable records. They are typically driven by a desire to use their investments as part of a commitment to lives that improve the world. The retail market for RI products is notably strong in the United States and Japan. In the United States, retail investors, along with religious organisations, were historically one of the driving forces of the RI movement as it evolved during the 1970s and 1980s.

Responsible retail investors usually invest in RI mutual funds (unit trusts) or, if wealthy, through separate accounts managed by private banks or trust offices. They tend to be passionate about their issues and hope for change in the world sooner rather than later. They may be strongly focused on a single issue – for example, environment, animal rights, labour or union relations, or nuclear disarmament. They may respond strongly to the headline news – for example, oil spills, obesity, child labour, bottled water or blood diamonds, or they may be generally concerned about the state of society and the environment. Those investing in mutual funds must adopt the general RI standards and engagement policies of the fund, while those working with bank trust officers or financial advisers can tailor their accounts to investments representing more focused and specific standards.

Their level of financial sophistication varies greatly. Because they invest on their own account, they can, if they wish, accept lower returns from the marketplace in the short term or even in the longer term without contending with complications arising from fiduciary issues. Many, if not most, are committed to obtaining a market-rate return. The choice is entirely in their hands.

According to the US Social Investment Forum, as of 2007 in the United States there were 260 RI mutual funds with $202 billion in assets.[32] These funds primarily serve retail RI investors and retirement savings plans (defined contribution pension plans) for individuals and to a lesser extent small institutional investors. In Europe, the number of RI retail funds in 2007 was 437, representing €49 billion in assets.[33]

In Japan, responsible investing began in the mid-1990s with the launch of a number of retail funds with environmental and sustainability themes.[34] As of 2007, there were some 34 RI funds in Japan with combined assets of approximately ¥58 billion, or approximately US$3 billion. As of 2010, institutional investors played a relatively minor role in the Japanese RI market.

Small institutional investors

Small institutions adopting a responsible investment philosophy tend to be mission-driven non-profits such as religious organisations, foundations, healthcare facilities and universities, or environmental, social services and human rights organisations. Those with endowments may choose to allocate all or some portion of their assets to responsible investment. Frequently, those with a defined contribution savings retirement plan for their employees offer an RI option. Their assets under management can range anywhere from a few million dollars up to several billion dollars.

The impetus for these organisations to engage in RI typically stems from their mission. Most are focused on a specific societal or environmental mission, so deciding that the management of their finances should be consistent with that mission can serve as a logical extension of their more general work. For example, the Robert Wood Johnson Foundation, which is devoted to healthcare issues, has a policy of not investing its endowment in the stock of tobacco companies.

During the 1970s and 1980s, the RI movement in the United States was led in large part by religious organisations. Many had long had prohibitions on investment in 'sin stocks', but starting in the early 1970s, many began incorporating into their investment policies and practices issues such as peace and militarism, environmental justice and equal rights for women and minorities. Religious organisations in particular pioneered the South Africa divestment movement during those years. Working through the Interfaith Center on Corporate Responsibility, they were leaders in the filing of shareholder resolutions and engagement with corporations. Among the religious organisations with comprehensive RI policies and practices as of 2010 were the American Friends Service Committee (Society of Friends), United Methodist Church, Evangelical Lutheran Church in America and the Presbyterian Church (USA).

The university and foundation worlds in the United States have generally speaking been slow to adopt RI practices. However, as of 2011, over 90 foundations had joined the More for Mission coalition and over 120 were members of the PRImakers Network, which promote respectively market-rate and below-market rate investments of foundations assets in mission-related ventures.

Large institutional investors (governmental pension funds and sovereign wealth funds)

Large institutional investors implementing RI polices and programmes are national, state or local pension funds, trade union pension plans or sovereign wealth funds. These funds are among the largest pools of assets in the world, ranging from the tens of billions to hundreds of billions of dollars in size. They tend to be invested in the full range of asset classes and are invested globally. In Europe, institutional investors tend to dominate the RI markets. For example, as of 2009 in France institutional investors accounted for 69% of the €50.7 billion RI market, down somewhat from 75% in 2008, according to the research firm Novethic.[35]

It has only been since the late 1990s that institutional investors, led particularly by those in Europe, have become a major factor in the responsible investment movement.[36] In the 1970s and 1980s, a substantial number of US state and municipal pension funds participated in the South Africa divestment movement protesting the apartheid legal system. At its peak in the late 1980s and early 1990s, those of the states of California, New York, New Jersey, Massachusetts and Wisconsin, among others, with hundreds of billions of dollars in assets, had policies limiting investments in companies doing business in South Africa.[37] With the dismantling of apartheid in 1994, these American pension funds pulled back from these RI activities and concentrated their activism primarily on corporate governance issues.[38]

With the growth of interest in sustainability in Europe starting in the mid-1990s, and the simultaneous privatisation of many industries previously in state hands, governments and institutional investors became increasingly interested in responsible investment. Governments, including those of the European Union, began promoting the concept of CSR and various national and local pension funds began adopting aspects of RI. For example, in the late 1990s and early 2000s the governments of the United Kingdom, Germany and Sweden, among others, adopted policies requiring pension funds to state whether they took societal and environmental considerations into account in their investment practices (see Fig. 3). At the European Union level, discussions around increased transparency by institutional investors on their RI practices were under way as of early 2009.[39]

Figure 3 **The adoption of disclosure regulations for pension funds**

Source: adapted from C. Louche and S. Lydenberg, 'Responsible Investment', in J.R. Boatright (ed.) *Finance Ethics: Critical Issues in Financial Theory and Practice* (Hoboken, NJ: Wiley, 2010): 408

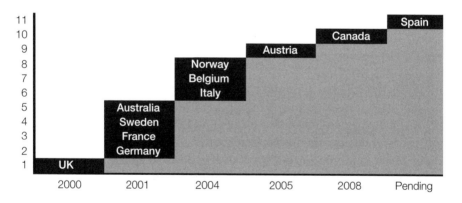

As a result, a number of large European pension plans have increasingly adopted specific RI practices. The Norwegian national pension fund has established specific criteria on weapons, human rights, bribery and the environment.[40] The Swedish national pension plans have varying policies, including several with weapons and human rights criteria.[41] The French FRR is planning to apply sustainability criteria across all asset classes.[42]

One of the most significant developments in the institutional investor field has been the creation of the Principles for Responsible Investment (PRI),[43] under the aegis of the United Nations Global Compact. Launched in 2006, the PRI had been endorsed by approximately 150 pension funds and other large institutional investors as of early 2009, including major pension funds from throughout the world.

By signing the PRI, they agree to implement six basic practices in their investing:

1. Incorporate ESG issues into their investment analysis and decision-making

2. Incorporate ESG issues into their ownership policies and practices

3. Seek ESG disclosure

4. Promote the PRI principles within the financial industry

5. Work cooperatively to implement the PRI principles

6. Report on progress in implementing the PRI principles

The PRI is particularly significant because it provides a forum to institutional investors for consensus building on best practices, collaborative action, engagement and the promotion of academic work in the field.

Both large and small institutional investors must deal with fiduciary issues and assure that they can provide adequate funds for their organisations, in the case of endowments, or adequate benefits for pensioners in the case of retirement plans. Fiduciary duties are legal obligations and subject to various interpretations of the law. The fact that investments have societal and environmental implications can pose dilemmas for fiduciaries because they are often told that they should simply ignore them. Yet it is increasingly clear that these implications affect the well-being of their beneficiaries in ways that have financial as well as societal consequences.

Asset managers

One can distinguish three primary types of asset managers within the RI space: money mangers who specialise in exclusively serving RI clientele; mainstream money managers who serve RI clients as well as those without RI concerns; and financial consultants and investment advisers who, although they don't management money directly themselves, can steer clients with RI concerns to those who do.

Responsible investment specialists

RI specialists are firms who apply RI policies and practices to all their assets. Some firms have a client base committed to using the full range of RI tools – standard-setting and engagement – in their portfolios, while others may use only engagement. Both are committed to the principle that all investments have societal and environmental implications and recognise their obligation to evaluate these implications and act – through divestment or engagement – when concerns rise to a certain level.

RI firms using both standard-setting and engagement for all clients – such as the US money management firms Trillium Asset Management and Walden Asset Management, and the US mutual fund firms Calvert, Pax World and Domini – tend to be relatively small, but command attention disproportionate to their size.

A number of large money managers and financial institutions in the United Kingdom have committed to applying responsible investment principles to all of their assets. These include: F&C Investments, which describes its commitment to responsible investment as 'fundamental to our global investment philosophy across all our funds'. In addition to managing some £3.4 billion in funds with RI criteria, it engages with corporate management on behalf of all its own funds, as well as providing an engagement service to others;[44] Hermes Asset Management – owned by its largest client, the BT Pension Scheme, Hermes describes itself as 'completely committed to responsible investment and the long-term approach that it entails'. Hermes manages some £27 billion in funds for a variety of clients, including BT, with an interest in responsible investment;[45] and Co-operative Bank, a part of the Co-operative Group. A large UK consumer cooperative organisation, the bank had £18 billion in unit trust assets under management as of September 2008 and analyses 'social, ethical, environmental and other company management issues (e.g., "fat cat" pay) across all the funds we manage'.[46]

One rapidly growing segment of the institutional RI market is Islamic investing. Investors in Islamic countries have become increasingly interested in developing and applying approaches based on the teachings of the Koran. As interpreted by most Sharia committees (local committees of clerics charged with applying the Koran's principles to daily life), the Islamic approach resembles traditional Christian RI screening in that, for example, it avoids companies producing alcohol. However, its interpretation of the Koran's condemnation of usury typically requires the screening out of financial services companies. According to a report by the Oliver Wyman consulting firm, Islamic finance will reach $1.6 trillion in assets by 2012, and as of year-end 2007 had $660 billion under management.[47]

Mainstream financial firms

As the market for responsible investment grows, as more and more clients demand the integration of societal and environmental factors into their investment processes, mainstream money management firms have opted to serve this client base. Firms such as State Street Global Advisors and Wellington Asset Management have RI specialists on staff who implement a variety of RI strategies for a range of clients.

These firms are generally client-driven and use standards-setting rather than engagement to implement their strategies. For example, they might manage accounts for state or city pension funds operating under legislative mandates to avoid tobacco companies or companies with operations in Sudan; they might also run funds for church groups that want to avoid 'sin stocks', develop a criterion for an animal rights organisations, and so on.

Such firms can be distinguished from their UK counterparts in the sense that they make no public assertion that they believe societal and environmental factors are an important investment consideration for their other assets. However, they are willing to incorporate these factors into stock selection at the guidance of their clients.

Initially, it was the South Africa divestment movement of the 1980s that prompted these mainstream firms to serve RI clients. As a broad variety of institutions increasingly take up one or another aspect of RI, it is rare to find a large money management firm that will not to accommodate these RI clients.

Financial advisers and consultants

With both asset owners and asset managers increasingly interested in RI, financial intermediaries who match up those seeking RI services with those offering them play an increasingly important role in the process. For institutional investors, these intermediaries are called financial consultants. For retail investors they are called financial or retail investment advisers.

On the retail side, in the US several financial adviser networks specialise in RI, including First Affirmative Financial Network and Progressive Asset Management.

As interest in RI has grown, financial consultants – who serve as gate-keepers for institutional investors, advising them on management practices and helping them implement their financial objectives – are increasingly recognising responsible investment as a legitimate discipline and advising interested clients on how best to enter the field. Among the major financial consulting firms with RI teams in place as of 2010 were Mercer, Cambridge Associates and Watson Wyatt.

Support services

A growing number of RI research firms and rating agencies, as well as in-house green teams within mainstream money management firms, have also emerged. These in-house teams and outside research firms play a crucial role in supporting the engagement between the financial community and corporate management on societal and environmental issues and are important intermediaries between companies and fund managers as they have gained legitimacy in their assessments of companies.

Research firms

To serve the growing need for data to implement RI strategies and tactics, a number of RI research organisations have sprung up. These organisations provide background data on the societal and environmental records of publicly traded companies, rating and ranking their performance. The information they provide is used primarily by institutional investors for investment decisions or shareholder engagement.

In 2007, the French consultancy Observatoire sur la Responsabilité Sociétale des Entreprises (ORSE) and Agence de l'Environnement et de la Maîtrise d'Energie (ADEME), in an update of a study originally published in 2001, surveyed the methodologies and services of 30 RI research and rating organisations.[48] Among the major firms are EIRIS (United Kingdom), GES Investment Services (Scandinavia), Jantzi/Sustainalytics (Canada/Netherlands), PIRC (United Kingdom), MSCI ESG Research and Screening, including KLD Research & Analytics (United States), SIRIS (Australia) and Vigeo (France).

Engagement services

Engagement with corporations to encourage positive change has been an important feature of RI since its inception. During the 1970s, engagement was in many senses the primary focus of RI in the United States. Since 2000 it has become increasingly an important part of the sustainable investment movement. Among organisations with a strong focus on engagement today are: F&C Investments, which has an engagement protocol called Responsible Engagement Overlay (reo®) that it applies to all its assets plus an additional £63 billion of funds managed by other investment institutions;[49] Principles for Responsible Investment Engagement Clearinghouse, which provides its institutional investors with a platform for sharing information on engagement activities and encourages collaborative engagement efforts;[50] and GES Investment Services Engagement Forum, which facilitates collaborative actions among northern European institutional investors, with a particular focus on encouraging companies to meet international norms on societal, environmental and governance issues.[51]

In addition, proxy voting advisory services, such as MSCI ESG Research and Screening and Glass Lewis in North America and PIRC in the United Kingdom, provide recommendations on how to vote on the numerous shareholder resolutions filed each year on corporate ESG issues.

RI associations and think-tanks

With the growth of RI, a number of associations facilitating networking, meetings and the promotion of best practices have evolved. These initiatives bring together professional practitioners, researchers and academics.

The most widespread model for these associations is the social investment forum, or SIF. As of 2010, the European SIF, now known as the European Sustainable Investment Forum (Eurosif) served as an umbrella for several European national SIFs – including Belgium, France, Germany, Italy, the Netherlands, Sweden and the United Kingdom – and was affiliated with four other SIFs in Asia, Australia, Canada and the United States.[52]

In addition to the Principles for Responsible Investment, other coalitions of investors organised around particular issues, such as the Carbon Disclosure Project and the Investors Network on Climate Risk, are also increasingly emerging and attracting widespread participation. In addition, organisations such as the Enhanced Analytics Initiative have been formed to influence money managers and fund owners to factor into their investment practices those negative and positive externalities relating to ESG that companies have created.

An increasing number of RI academic research initiatives are also under way, including the Moskowitz Research Program at the Center for Responsible Business at the University of California, Berkeley; the Sustainable Investment Research Platform, hosted by the Umeå School of Business in Sweden and sponsored by the Mistra Research Programme on Sustainable Investments; the European Centre for Corporate Engagement, a joint initiative of Maastricht University and RSM Erasmus University in the Netherlands; and the Initiative for Responsible Investment at the Harvard Kennedy School's Hauser Center for Nonprofit Organizations.

The future of responsible investment

As it develops and matures the field of responsible investment has become more complex. Clients raise an increasing number of societal and environmental concerns. Corporations simultaneously create challenges and make efforts to solve an increasing number of problems. The number of players serving these clients and engaging with these corporations expands and these players apply a variety of new and interesting approaches. Government's role in the interaction between corporations and society evolves, sometimes putting into private hands businesses that were previously state-owned – and at other times stepping in to impose greater regulations on corporations' activities.

Moreover, the concept of responsible investment is finding traction in other asset classes beyond the stocks of publicly traded corporations. It is increasingly clear that responsible investment in real estate can improve the quality of our daily lives for generations to come. Venture capital that promotes clean technology and alternative energy has the potential to revolutionise our energy systems. Investments in sustainable forestry, organic agriculture and other commodities can help undo damage already done to our environment.

This extension of responsible investment to a growing world of clients, corporations, mainstream financial players and differing asset classes creates a complex kaleidoscope through which this emerging world can be viewed. It forms a complex picture with many moving parts. This diversity allows for considerable flexibility in RI's application, but it can also pose difficult choices and a variety of dilemmas for those practising this discipline every day.

Our hope in presenting the following 12 case studies is that we can convey some of these complexities in all their richness, along with some of the dilemmas that arise in contending with the variety that characterises this world. For it is in RI's diversity, in its ability to serve a broad range of clients and address a multiplicity of issues, that its strengths lie. This variety and complexity may make the daily practice of RI a more complicated exercise than that of the relatively simple task that the mainstream investment community of today sets itself – matching or beating a financial benchmark. However, it is precisely this variety and complexity that allow responsible investors to understand and contend with the societal and environmental implications of their choices.

Case 1

Types of responsible investor

- How customised should a responsible investment product be?
- How do business models influence responsible investment product offerings?
- Is there a single optimal responsible investment portfolio?

The case

You are a conventional money manager and you have just become interested in the responsible investment (RI) market. You believe that RI is a growth market and you would like to enter it. You issue a press release announcing that you are now offering RI services. Your announcement receives wide coverage and the telephones start ringing. For next Monday you have set up meetings with four potential new clients. The first is Mary, a single working mother with three children, passionate about a whole range of societal and environmental issues, with a modest amount of money to invest. The second is Christian, an independently wealthy man in his sixties, who has just heard about RI and is intrigued because, for the first time in his life, he is becoming concerned about environmental issues. For him, incorporating environmental concerns in his investment portfolio seems like a 'nice idea', but it seems to you that this may be no more than an idle thought to him. The third is Paula, the chief financial officer for a small church that has an endowment to invest. The members of its investment committee are adamant that its endowment should be invested according to principles of fairness and societal justice. The church depends on revenues from its endowment for its operating expenses. The fourth is Rob, the head of the board of trustees for a large pension fund that is being pressured by its retirees not to invest in companies that manufacture landmines. He is not at all interested in RI or the landmine issue, but wants to discuss various options with you.

How would you prepare for this meeting . . . ?

Dilemma for the responsible investor

This case highlights dilemmas that arise because investors come to RI with (1) differing levels of sophistication about the field, (2) a multiplicity of specific societal and environmental concerns, and (3) differing conceptions about how they want to become involved in the RI process. This variety poses challenges to managers serving these clients.[53]

Because investors have different risk tolerances with respect to financial returns, traditional money managers regularly encounter similar situations. They assess both their clients' tolerance for risk and their financial needs. Based on these assessments, they then construct an appropriate portfolio. Given certain risk and reward characteristics, managers can in theory identify an optimal portfolio for each client.

In addition to making the necessary assessments of clients' risk tolerances and financial needs, RI professionals also seek to understand and incorporate into the investment process their clients' societal and environmental concerns. Clients can differ substantially on the issues about which they care and can possess substantially different levels of understanding of, and sophistication about, these issues. One client may be only vaguely informed on one issue, and therefore look to the manager for guidance, but on a second issue may have an understanding that far surpasses that of the financial professional.

Furthermore, decisions about how to incorporate societal and environmental concerns into security selection can have implications for portfolio construction. Money managers and their clients need to be aware of these implications. Imposing excessively stringent societal or environmental concerns can curtail diversification options. Less stringent concerns may be

The case in perspective

This case focuses on the variety of societal and environmental concerns that responsible investors bring to the investment process and the consequent need for those serving these investors to create a variety of customised products that take these concerns into account. Because this variety is essentially without limit, those serving this clientele must decide when and how to limit their customisation of products in ways that are practical and yet serve their clients' needs.

That responsible investment reflects in its products and services the societal and environmental concerns of investors is one of its most important characteristics. These products and services legitimise the principle that investment decisions have societal and environmental implications.

Responsible investors may mistakenly believe, however, that because their products and services can, to some extent, reflect these concerns, that ➔

they can also address them – or even solve them – without limit. The struggles of those serving these clients to confront these limits, while still addressing their clients' major concerns, enrich the range of investment products and services that the financial community provides.

able to be substantively addressed without such limitations. However, for example, if a client requests a manager to avoid all companies that have no women on their boards of directors or all companies that contract with overseas vendors that might have labour issues, an investment professional should explain that so radical an approach would limit investment options. The range of issues clients care about, and the degree to which they care, are as varied as the number of ways corporations interact with society and the variety of personal concerns. Their implications for portfolio construction need to be understood.

When dealing with institutions, money managers may also need to contend with situations in which opinions about societal and environmental issues vary among the decision-makers. A board of trustees can be split on which issues should be incorporated into its investment policy, or to what degree. In addition, responsible investment managers may have personal opinions about specific issues that differ from those of their clients. Finally, responsible investors are no different from other investors in the sense that some want to be involved in the daily decisions and others just want the manager to do the job.

Because of these variables in interests and sophistication, managers can face complex choices between customisation and standardisation of the services or products offered to clients. Decisions here are also affected by the managers' business models or commitment to offering specialised client services.

Approaches available to the responsible investor

One of the key themes raised in this case is that of customisation and flexibility. Because RI clients differ in their interests and sophistication and investment managers differ in their business models and commitments to customised services, finding the appropriate match between clients' wishes and managers' approaches can be a challenge. Customisation can take a variety of different forms. A manager can:

- Offer multiple products that serve a range of investors with a variety of specialised RI concerns

- Offer a single product with one set of clear standards that can be said to embody the manager's RI concerns

- Encourage clients to express the nature of their concerns and the specific companies they are and are not comfortable holding in their portfolios

- Encourage clients to rely on their judgement on how best to select companies on societal and environmental grounds

If an RI money management firm offers the full range of products customised to all clients' concerns, it runs the risk of having no two accounts the same. This approach has the advantage of enabling the manager to serve all clients, but it can be time-consuming and expensive. By contrast, a single product that addresses a single set of issues makes a clear statement – creates a kind of public brand – with which investors can identify, particularly if they are not sophisticated about its particulars. The disadvantage of the latter approach is that it cannot serve investors with strongly held specific concerns that differ from those on which the manager has chosen to focus.

The nature of an RI manager's business model will also influence the available choices. For example, because of the economics of the mutual fund business (it is expensive to launch and maintain a mutual fund) managers of mutual funds offer a limited number of choices. On the other hand, a bank trust officer with many separately managed accounts may have the flexibility to tailor portfolios more closely to clients' societal and environmental concerns.

Whatever the choices particular RI managers happen to make, in the end a substantial number and variety of RI products and services are inevitably being offered. The variety and levels of concerns about societal and environmental issues raised on behalf of RI investors and the number of different business models adopted by RI managers essentially guarantee that this variety will be a part of the RI world.

Variable factors

Responses to this case may differ depending on a number of factors, such as the manger's resources and the interest of the client in particular RI issues and in taking an active role in the RI process.

- What resources does the money manager have available?
  - If the manager is with a large firm, is it willing to devote resources to creating separately managed, customised accounts?
  - If the manager's firm offers mutual funds, is it willing to go to the expense of creating separate publicly traded mutual funds?
- What is the manager's business model when it comes to responsible investing?
  - Is the manager primarily committed to RI or is this a sideline business?
- What is the client's level of concern and sophistication about the issues?
  - To what degree does the client have strong views about particular RI issues?
  - How knowledgeable and sophisticated is the client about these issues?
- What are the diversification implications of the RI concerns?
  - Are the client's guidelines likely to affect portfolio construction?

Recommendations

Responsible investors faced with similar dilemmas may want to:

- Understand the full range of clients' societal and environmental motivations and concerns
- Explain clearly to clients the differing aspects of RI research and engagement policies and practices
- Evaluate with clients the potential portfolio construction implications of the different RI options
- Ensure that the range of RI products offered is consistent with the business model and resources of the manager's organisation
- Be sure to be able to deliver what is promised on both the societal/environmental and the financial sides of the equation

The RI pendulum

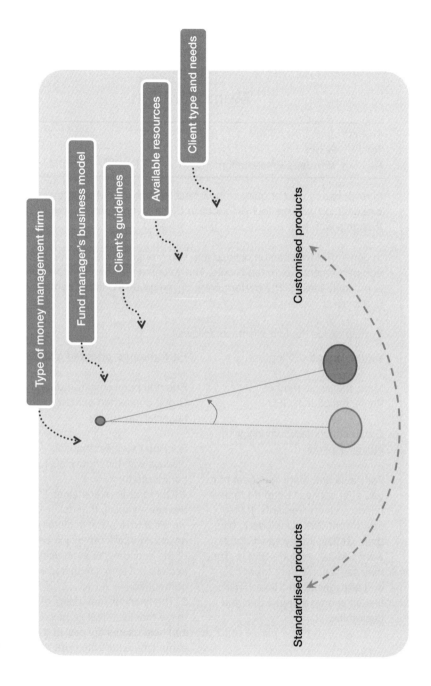

Responses from practitioners

Value-driven and performance-driven investors

Memorandum

From: Jean
To: Robin
Re: Monday's Presentation

In response to your enquiry on how best to approach Monday's presentation with the four prospective clients, let me share some of my thoughts with you.

In general, clients can be divided into two groups: values-oriented clients and performance-oriented clients. The values-driven group's concern is primarily ethical. The performance-driven group's primary concern is financial.

I would divide the four clients as follows:

Values-oriented clients | **Performance-oriented clients**

Mary, the single mother
Paula, the church CFO
↓
Group concerned primarily with ethical aspects

For Paula and Mary I suggest that you offer services adapted to their needs and requirements. If they say 'I want GMOs included', give them GMOs. If they say 'I don't want GMOs', exclude GMOs. They may say 'I don't want tobacco' or 'I am against abortion'. That's fine. In short, I suggest that you customise.

Rob, the pension fund trustee
Group concerned primarily with financial materiality
↓
For Rob I suggest that you discuss with him how ESG factors contribute to:
1. Our long-term sustainability themes – explain that as investors we want to support societally and environmentally promising sectors of the economy, or even risky sectors but with a high societal contribution.
2. The financial materiality of these themes – that is, how ESG can help us identify and at the same time anticipate potentially profitable business models.

- page 1 of 2 -

For Christian, it is not clear how deeply his commitment to environmental issues runs. I suggest that you pose a series of hypothetical questions to him. Give him examples of companies with environmental strengths and concerns and see how deeply he feels about each. Ask him what kinds of financial trade-off, if any, he might be willing to make to ensure that certain types of company are excluded from, or included in, his portfolio.

Remember these important points:

- In general, our clients have the right to be demanding and express their own values in their investments, although the situation is more complicated for pension funds.
- We should not systematically oppose either the values approach or the performance-driven approach, but allow room for both according to the nature of our clients.

> 66 I would really sit down and try to consider all the issues, with each prospect. Then it's a trade-off. So when you introduce the ESG [environmental, societal and governmental] strategies you need to figure out where the commonalities are, but as you meet with each individual you need to figure out what all the issues are, get them on the table, and determine what is more material and try to get a feeling for their social mission relative to their current investment agenda. 99

Four clients, four different approaches

Memorandum

From: Neil
To: Julie
Re: Monday's Presentation

Your presentations on Monday pose some interesting challenges. Because these four potential clients are very different from one another, I recommend that you take a different approach with each. In particular, be certain that each client fully understands both the financial and the societal implications of their decisions. Here are the approaches I would suggest for each client.

Mary is particularly motivated by RI concerns. I would recommend that you stress the financial implications of the various approaches she might take.

Christian is presumably already sophisticated about the financial markets and the financial implications of his investments. I would recommend focusing on explanations of RI practices and how they fit into conventional investment approaches and his own current investment strategies. We need to be quite clear on what his financial objectives are.

For Paula, her church is likely to already have detailed RI guidelines, as well as financial objectives. You should determine with her if any of our current products fit these objectives. If there isn't a match, we can consider coming up with a customised product, or they can determine how comfortable they are with those of our off-the-shelf products that most closely match their guidelines and objectives. An off-the-shelf product would be more efficient and less costly for us.

For Rob, the situation is more complicated. The pension fund already has certain requests from its members, so the question isn't just what products we have or what we could recommend. We can easily screen just for landmines, but we need to assess the implications of the full range of screens requested. This is further complicated by the fiduciary regulations under which the pension plan operates. Fiduciary regulations vary from country to country, so you must be quite clear as to what fiduciary implications this request has for our products or any customised products we might create for this client. You should be aware that the trustees of the pension fund may be very cautious and reluctant to take up requests not to invest in landmines.

One similar presentation for all four clients

Memorandum

From: Jack
To: Oliver
Re: Monday's Presentation

I would recommend that you make a similar presentation to all four clients. The presentation should introduce the concept of responsible investment, provide background on its history, sketch out various approaches to RI, and give an overview of our RI products.

Then you should have a detailed discussion with each client, in which you:

- Explore thoroughly the depth of the societal and environmental concerns of each client.
- Determine the financial risk tolerance of each client and understand how, if at all, their societal and environmental concerns might affect this risk tolerance.
- Make certain that each client understands the financial implications of various approaches. With Christian and Rob, be certain they understand the financial effects their concerns might have. With Mary and Paula, be certain they understand the 'social return' they can expect.

> 66 The financial officer for a small church is actually easier because they have certain principles, so it's just about how we come up with the implementation process – it's easy, because they already have a policy, e.g. on alcohol. So we just refine the screening process. The pension fund is the most difficult because it doesn't have its own value drivers in this space, so we have to first educate them on what RI is about. 99

In the news

The following examples illustrate the degree of difference in approach that can exist within various types of responsible investors. Here we have chosen two foundations and two pensions funds that have taken differing approaches to the RI issues they wish to incorporate into their investment policies and practices.

The Robert Wood Johnson Foundation (RWJF) excludes tobacco companies from its portfolio. RWJF was founded by the head of Johnson & Johnson in 1936 and has a mission to improve health and healthcare in the United States.[54] In 1990, consistent with that mission, RWJF adopted policies prohibiting investment in tobacco companies. At that time, the foundation had allocated nearly half a billion dollars to various organisations in an effort to reduce tobacco use. It did not feel that this campaign could be effective if the foundation continued to support tobacco companies through its endowment's investments.[55]

The Jessie Smith Noyes Foundation, established in 1947, supports grassroots organisations working towards the creation of a sustainable society. The foundation's investment priorities are shaped by this mission. It uses both exclusionary and inclusionary standards to identify investments that align with the foundation's mission in four areas: toxic emissions, extractive industries and environmental justice; sustainable agriculture and food systems; reproductive health and rights; and a sustainable and socially just society. For example, it excludes companies that produce synthetic fertilisers and pesticides, but makes a point of including companies that produce, distribute or sell organic products.[56]

The Ohio Public Employees Retirement System (OPERS) engages with companies operating in Iran and the Sudan, and reserves the right to divest if the companies are not responsive. In response to a proposed state law that would have mandated public investors in Ohio to divest from companies conducting specific types of business in Iran and the Sudan, OPERS adopted its own policy. Concerned about its fiduciary responsibilities, as well as about the risks of investing in companies with ties to these countries, OPERS elected to require its portfolio managers to identify and engage with at-risk companies operating in Iran and the Sudan. Investments in companies that do not move to address these risks may be divested from the portfolio, if there are other 'comparable investments offering similar quality, returns, and safety'.[57]

Stichting Pensioenfonds ABP (ABP) excludes companies from its portfolio that are involved in businesses that conflict with international law, primarily those relating to military products. ABP is the Netherlands' pension fund for government and education employees. ABP does not invest in companies that deal in products or services prohibited by Dutch or international law, and so excludes from its portfolio companies involved in the production of landmines, cluster bombs and chemical or biological weapons.[58]

Cases for comparison

Compare and contrast this case with Case 8 'Relativity of responsible investment standards'.

Case 2
Ethics and facts

- Is there a legitimate role for ethics in responsible investment?
- What is the relationship between ethical concerns, hard facts and emotional responses?
- What role should ethics play when facts are not entirely clear?

The case

You are a money manager with a long-term RI client very concerned about human rights issues. The client, Mr Jones, calls you and explains that he has just read an article in the paper that says that one of the companies you have put in his portfolio – Company X, a manufacturer of electronic games – has been contracting with suppliers in China that have horribly abusive labour conditions. Your client is passionate about human rights issues and impatient with any company that has a record of abuse in this area. He is outraged by the report and wants you to sell the stock immediately. He has a large individual account with you, which you want to keep. The situation is complicated, however. The company denies the allegations and points to its otherwise excellent corporate social responsibility (CSR) record. It is unclear what all the facts in the situation are. Human rights groups can't tell you if other electronic games companies in your portfolio have the same issues with their contractors, but they say that working conditions are a general problem with the electronics games industry as a whole.

What next steps do you take . . . ?

Dilemma for the responsible investor

This case highlights the dilemmas that can arise when responsible investors raise ethical issues with regards to their investments. The mainstream investment community, which generally favours an unemotional, strictly rational approach to investment decision-making, tends to view ethical considerations as 'emotional' or 'irrational' responses.

From the mainstream financial community's point of view, hardheaded consideration of only the financial facts is the best way to pick stocks. When, in its view, emotions prevail, investors are all too likely to fall in love with a stock that has done well for them, or hate one that hasn't, clouding their abilities to realistically assess a company's financial prospects. As behavioural economists have pointed out, emotions often do come into play in the investment process.[59] In the view of the mainstream, for the markets to function efficiently and to allocate assets appropriately, investors should strive to be as unemotional about stock picking as possible.

Because responsible investors may express their views about the ethical implications of investment decisions strongly, their reactions may be viewed by the mainstream with suspicion.

Responsible investors tend to view their role in the investment process differently. They acknowledge that investment decisions have societal and environmental implications. They view assessing the ethical implications of these decisions as a legitimate part of their role as investors. They do not wish to profit from unethical practices. They understand that unethical corporate decisions can damage the quality of the life we lead and our society as a whole.

From the responsible investor's point of view, therefore, expressing ethical opinions about corporate behaviour is perfectly reasonable. Raising these issues can help corporations confront and address

The case in perspective

This case focuses on the legitimacy of ethical responses to investment decisions. Responsible investment acknowledges that under certain circumstances it is appropriate to incorporate ethical considerations into the investment process. Mainstream finance tends to downplay ethical responses, viewing them as similar to emotional or irrational responses inappropriate in investments.

Because investment is not an isolated activity and cannot be separated from the daily conduct of corporations in society, it often has ethical implications. Consequently, responsible investors often have strong responses to the ethical aspects of that conduct. However, the circumstances surrounding controversial situations, while sometimes clear,

→

unethical behaviour. It calls the attention of public officials and policy-makers to regulatory frameworks that may need change. It helps investors avoid risks to a company's reputation or helps a firm avoid litigation further down the road.

are not always so. Determining what responses are appropriate and what actions can achieve desired ends in these situations can pose substantial challenges.

Once questions of ethics have been introduced, however, they can lead to complicated discussions about the particulars of the corporate conduct in question, the facts relating to a particular company's conduct, the practices that are generally followed within its industry, the appropriateness of the public-policy framework within which this conduct is taking place,

It is in confronting these challenges that responsible investors help develop and deepen an orderly dialogue between corporations and society around ethical issues and help remind us of the ethical implications of investment decisions.

and the ability of the general public, along with responsible investment professionals, to obtain the facts.

The dilemma for responsible investors is how, in their communications with companies, investors and the public, to appropriately manage the ethical dimensions of communications about corporate misconduct. These responses, if not held in check, can complicate assessments of the facts and the subtleties of complex situations.

The challenge for responsible investors is to incorporate the ethical aspects of their assessments in their communications with corporations and the public in such a way as to give them their due, but at the same time avoid precipitous action or frivolous debate.

Approaches available to the responsible investor

One of the key themes raised in this case is the role of communications in dealing with ethical issues in the investment process. Responsible investors can strive to use thorough and thoughtful communications to ensure that ethical issues are incorporated appropriately into the investment process. Such communications include:

- Clear articulation of the ethical standards in question, their importance and the grounds on which they are generally recognised as important

- Comprehensive statements about the company's historical, as well as current, behaviour relative to those standards

- Thorough explanations to both clients and corporations about the research methodologies and findings used

- Consideration of opportunities to take public positions on the legal or public policy framework within which the issue is being raised

Such communications can be crucial in clarifying the importance of ethical issues, understanding and managing the strong responses they often evoke, and encouraging reasonable action.

Many ethical standards of concern to responsible investors are based on international norms such as the United Nations Universal Declaration of Human Rights and the International Labour Organization's Declaration of Fundamental Rights at Work. Clear communications about the existence of these widely recognised norms and codes of conduct can help to establish that a framework within which corporate behaviour can be evaluated exists and that responsible investors' concerns are not a matter of idiosyncratic or personal opinion.

Although a particular controversy may involve highly controversial corporate behaviour, responsible investors will want to demonstrate that they understand the company's historical record on the issue in question and communicate with the company within the context of that understanding. For example, acknowledging that the company has had a previously strong record on corporate social responsibility can help ensure that the investor's response to a particular situation is tempered by related historical facts. It is also important for responsible investors to communicate their understanding of whether the problem is isolated or widespread, whether the company has the necessary systems in place to remedy the problem and prevent it from recurring, and whether company is being collaborative or confrontational in its responses to the situation.

By publicly communicating the details of their research methodologies, responsible investors can also help establish the legitimacy of their concerns. These methodologies can include assessments of an issue at the sector level, establishment of a base line for comparison with similar firms, determination of best practices for a given industry, fact-finding at a local level and informed dialogue with corporate representatives. Communicating the details of a thorough, well-documented research methodology can go a long way towards assuring various parties that judgements made are generally fair and comprehensive.

Finally, responsible investors will want to be aware of opportunities that a controversy may create for them to engage in public dialogue on a particular issue and for participation in broader debate about public policy. Such debates may be hotly contested, but they can ultimately help establish broad consensus on particular issues.

Variable factors

Responses to this case may differ depending on a number of factors, such as the client, the source of information, the response of the company itself and the general practices of the industry within which it operates. In such cases, asking questions such as the following can be helpful.

- How knowledgeable is the client?
    - How much does the client understand about the issue involved? Would the client benefit from further communication and education on the issue?
    - How well does the client understand the manager's methodology? Would the client benefit from further education on the manager's standard-setting methodology?
- How trustworthy is information provided in the article?
    - Is this a case of a journalist just seeking headlines?
    - Is the article based on sound research and reliable sources?
- How is the company responding and reacting to the press article?
    - Is the company silent on this issue?
    - Is the company willing to cooperate with other organisations to find out what has really happened or is the company defensive or resistant to further questions?
- Is this a company-specific issue or sector-wide issue?
    - How frequently do such concerns arise within the company's sector?
    - How are the other companies within the same sector reacting?

Recommendations

As you approach dilemmas of this sort, you may want to:

- Build communications into the various aspects of the decision-making process with clients and corporations

- Be certain to have all the facts and figures and understand their significance and context

- Understand the motivations and positions taken by the full range of stakeholders

- Assure that current research methodologies and the key performance indicators that are used are sound

- Give the company adequate time to respond to any requests for further information or explanation

Recognising ethics

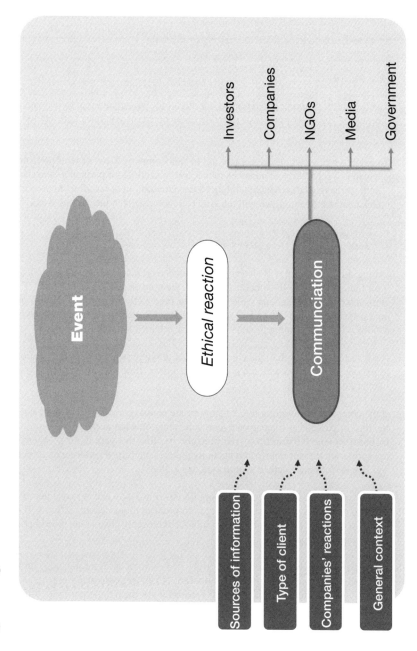

Responses from practitioners

Dialogue and engagement

Dear Mr Jones,

Thank you for bringing this report to our attention. We will attempt to respond as thoroughly and appropriately as possible to the concerns you raised during our phone conversation.

To begin with, we would advise you not to sell Company X out of your portfolio – which is a separately managed account customised to your particular concerns – until we have a better understanding of the situation. In order for us to properly understand the situation, we will take the following steps in the coming week:

- Contact the company
- Seek information from other sources

During this period, we will also conduct further research, analyse any new information, and only then make a decision. We may decide to exclude the company if there are serious violations of our fund's criteria or to put the company on a watch list. In the latter case, we would expect the company to take concrete actions within the coming months if it is to be retained as an investment.

If our research shows that the company is not going to take the necessary actions to remedy the problem within a reasonable length of time, the company will in all likelihood be excluded.

If the company is unwilling to take appropriate action, we will wait at least six months, or perhaps as long as one year, depending on what needs to be done, before removing the firm from your portfolio. If, after this period, the company has not shown significant improvements in its policies, monitoring system of suppliers and key performance data, it would then be excluded.

As this case involves changes at the supplier level, we may need to consider giving a period of up to a year to the company to implement the necessary reforms. It is always a long and difficult process to improve the supply chain. One should not expect rapid changes.

If this turns out to be is a particularly tricky and difficult case, we may also turn to our advisory board for further consultation. This board consists of experts in various fields, including human rights. We would ask them to help us with this decision.

- page 1 of 2 -

We might also consider making a visit to the supplier's site as it is the best and most certain means of really understanding what is happening.

With regard to the other companies from the same sector, we may contact them if the issue turns out to be particularly serious and could affect our assessment of their corporate social accountability records as well.

In this type of situation we generally prefer dialogue and engagement with the companies in question. We consider these dialogues not only a means to gain a better understanding of the company's policies and practices, but also a means of having a positive impact on companies' behaviour.

Please let us know if you wish to discuss this issue further. In any event, we will do everything in our power to get a better understanding of the situation to ensure that the company responds appropriately. If it does not, and if the concerns turn out to be substantive and severe, we will in all probability exclude the company from your portfolio.

Thank you for your cooperation.

Sincerely,

Your Money Manager

> 66 It's an opportunity to have a pretty detailed discussion with the client about what is possible to know and what's not always possible to know and how the client wants to deal with unknowns. 99

Communicate the methodology clearly

Dear Mr Jones,

Thank you for calling us and raising this concern about human rights issues. In such situations we believe that it is essential to seek a better understanding of what has happened, as well as follow up on the views and opinions of various stakeholders. Only then can appropriate action be taken.

As we proceed, and in order to avoid any misunderstandings, it is also essential that we communicate clearly with you about the methodology we use in such situations. First, we are operating under the assumption that, if the company is already in our portfolio, it has one of the best overall performance records of companies in its sector on societal and environmental issues. Second, it is very important to understand the context in which these criticisms have arisen and try to compare this specific company to the rest of the sector with regard to the issue in question.

Therefore, and in response to your concern, we would like to take the following three steps.

Step A: Schedule a meeting with you

The objective of this meeting is to reach a common understanding on what needs to be done before making a decision on whether to exclude the company or not. It is, at this preliminary stage, important to keep the company in the portfolio until we know more about what has happened and the allegations. Three main topics will be discussed during our meeting:

1. The importance of conducting further investigations and analysis before action can be taken.
2. The reasons for which the company is in the portfolio, and why it is the best in its sector and a top performer in corporate social responsibility.
3. Next steps that can be taken.

Step B: Conduct further research

The objective of this second step is to obtain a better understanding of the criticisms of the company and evaluate the opinions of various stakeholders. This step will involve:

- Obtaining the views and opinions of several stakeholders including the media.
- Contacting the company for its reactions and to assure that proper policies and procedures are in place at the firm to prevent this type of problem from arising again with its suppliers.
- Understanding the context within which the criticisms arose and comparing this company's policies and practices with those of others in its sector.

- page 1 of 2 -

Step C: Come to a decision

The objective of this third step is to come to a decision and discuss it with you. It involves:

- Analysis of all the information gathered.
- Follow-up meetings or phone calls with you to discuss and explain the decision.
- Ongoing monitoring of the company on this issue.

Thank you for sharing this concern with us. We will do our best to respond appropriately.

Sincerely,

Your Money Manager

> 66 We would seek as much information as possible from the media. We would always speak to the company directly. We would then look at other third parties that might be able to give us information such as NGOs. There might be other local research organisations. 99

> 66 We do not make decisions based on single issues but on the overall performance of the company. This will be part of the assessment but does not represent the full assessment. A single issue is a problem but it remains a single issue. It is important to ask if it's a systematic problem or just a single incident. 99

In the news

Companies dealing with difficult issues are often subject to cycles of positive and negative publicity. The following two cases illustrate the challenges of assessing corporations in times of controversy and the internal struggles that companies face when dealing with controversies and negative publicity.

Apple has been the subject of various controversies around working conditions at the vendors that manufacture its products. In mid-2006, allegations began to emerge of labour rights abuses and questionable working conditions at Foxconn, one of Apple Computer's iPod manufacturers in China.[60] Apple launched an investigation and found that Foxconn was in compliance with the majority of its Supplier Code of Conduct. However, they did find evidence of overcrowding in off-site workers' dormitories, an overly complex payroll and incentive structure, overtime violations and incidents of harsh treatment of workers.[61]

Foxconn agreed to make changes as a result of Apple's audit. Apple started a training programme for a 'no tolerance policy' regarding harsh treatment, hired an independent firm to assist in workplace audits and joined the Electronics Industry Code of Conduct Implementation Group. However, a May 2008 report issued by the Centre for Research on Multinational Corporations found that Apple had refused to provide substantive information on its auditing efforts.[62] Then in early 2010, a series of suicides at Foxconn's manufacturing plant in Shenzhen made Apple, along with other high-tech hardware companies, once again the focus of international controversy.[63] In June 2010, in response to these controversies, Foxconn agreed to raise the pay for all workers in its Chinese manufacturing plants by 30%.[64]

Criticised at some times and praised at others, Gap Inc. has worked to improve the working conditions at its suppliers' factories for over a decade. The anti-sweatshops movement of the 1990s targeted Gap, along with other companies such as Nike, for alleged complicity in child labour and other human rights violations by its suppliers around the world. Gap's efforts to address these issues included the creation of a Code of Vendor Conduct, formation of a compliance team to ensure adherence to the code, cancellation of contracts with non-compliant suppliers and issuance of public CSR reports heavily emphasising supply-chain management and performance. In the mid-2000s, Gap issued a series of annual CSR reports with ground-breaking documentation on the specifics of its suppliers, their level of compliance and the company's progress in improving performance – for which it received wide praise.

Then in 2007, an undercover investigation alleged discovery of child labour at a Gap subcontractor in India that was employing children as young as ten years. After an internal investigation into the matter, Gap cancelled the order that the supplier was working on, prohibited the subcontractor from receiving future Gap contracts, put the India-based vendor on probation, and met with all of its north Indian vendors to re-emphasise its no tolerance policy regarding child labour. To prevent future incidents, Gap improved its handwork supply-chain tracking system, defined preferred methods of handwork production, enhanced monitoring capabilities by increasing partnerships with local and global civil society organisations, and took steps to raise awareness of child labour at local, industry and national levels.[65]

Cases for comparison

Compare and contrast this case with Case 7 'Public versus private partnerships for engagement' and Case 9 'Incomplete societal and environmental data'.

Case 3
Influence through voice and exit

- How can responsible investors best influence a company's behaviour?
- When and how does selling a company's stocks become the best option?
- When and how does entering into dialogue with a company become the best option?

The case

You are a money manager with a client that is a large environmental foundation investing according to responsible investment principles. This is a major environmental foundation with assets of several hundred million euros. It is well known and quite influential. The chairman of the foundation, Mr Thomas, contacts you and says the foundation's board is unhappy with the environmental records of two chemical companies in your RI portfolio. It wants to influence both companies to improve their environmental practices. Company A is very willing to talk to you and the foundation about its problems, but Company B is not. As best as you can determine, both companies are moving slowly towards improvements, but not fast enough at the moment to satisfy the foundation. The foundation is willing either to sell its stock or to enter into dialogue with the companies

What steps would you recommend that the foundation take . . . ?

Dilemma for the responsible investor

This case highlights dilemmas that arise when a responsible investor wishing to influence a company's behaviour must choose between selling its stocks in public protest and remaining a stockowner who can engage the firm in dialogue. Each approach has its particular strengths, but also its limitations.

A number of mainstream money managers adopt an activist approach to investing, engaging with management on companies' financial performance. Historically, the 'Wall Street Walk' was the rule in the financial community – if you didn't like the way a company was being managed, you walked, that is, you sold its stock and bought shares in another, better-run, firm. Selling stock communicates to management your dissatisfaction through market mechanisms. If enough investors sell a company's stock, they drive its price down. Sometimes known as the 'discipline of the market', selling stock can be severe punishment indeed if it drives a stock down far enough to make the company a takeover target.

Since the 1980s, a number of institutional investors, including pension funds and hedge funds, and even wealthy individuals, have taken a different approach. When dissatisfied with management, instead of walking away, they have raised their voices, communicating their concerns to management about how their company is being run. Their dissatisfaction usually centres on poor stock performance. These activist stockowners typically urge management to adopt new strategies – for example, selling off a line of business – or even seek seats on the company's board of directors. They sometimes act together to pool their ownership and interests; and they can be quite effective in obtaining the change they seek.

The case in perspective

This case focuses on communications as a means of influencing corporate behaviour. Societal and environmental concerns – positive or negative – rise at times to levels that prompt responsible investors to communicate them to the corporations in which they hold shares. These communications heighten management's awareness of the depth and breadth of society's concerns. Various practices of mainstream finance today (e.g. index investing, high frequency and program trading) are so anonymous, mechanical and rapid that they no longer send clear market signals to corporate managers, depriving them of important inputs and depriving investors of one of their important responsibilities.

It is not always obvious which form of communication – dialogue, sale of stock, or a combination of the two – will be most effective in

→

Responsible investors similarly seek to influence corporate management, urging such changes as improvements in environmental management systems or vendor labour standards. They often argue that in the long term a connection exists between practices such as sound environmental management, strict overseas labour standards, or the like, and the company's long-term profitability and stock performance. As stockowners, however, they rarely hold enough shares individually or collectively to influence stock prices substantially. They face a challenge then as to what leverage they have in compelling management to listen to their concerns and take meaningful action.

Is ongoing dialogue the most effective route for small shareholders to communicate effectively with management on societal and environmental issues? Will selling the company's stock be of sufficient concern to management that it will change its ways? After selling one's stock, shareholders typically lose their ability to continue dialogue with management. But dialogue on these issues without the threat of divestment can be ineffective as well. What is the most effective approach?

communicating these concerns. In identifying effective means of communication, many specifics come into play, including industry norms, corporate cultures, individual personalities and a host of other factors. Understanding these specifics can be time-consuming and confusing and responsible investors frequently face dilemmas about how most effectively to approach companies with their concerns.

Properly handled, these communications can bring about beneficial change, demonstrating both the need for and the legitimacy of communications on societal and environmental issues. It is by continuing to struggle with the ongoing challenges inherent in these communications that responsible investors help keep corporate managers, as well as other shareowners and stakeholders, attuned to important societal and environmental challenges.

Approaches available to the responsible investor

One of the key themes raised in this case is influence. By seeking to influence companies to improve their societal and environmental performance, responsible investors generally hope to align corporate behaviour with broader objectives of society while generating better long-term financial returns .

The economist Albert Hirschman[66] has pointed out that members of any organisation – be they non-profit organisations, political parties or corporations – have two choices when they believe their organisation needs to improve its behaviour: they can communicate their dissatisfaction (what he calls 'voice') or they can leave the organisation (what he calls 'exit'). Voice without the possibility, or threat, of exit is weak, but exit undercuts the ability to raise one's voice.

Responsible investors, dissatisfied with a company in which they are invested and seeking change, can use voice and exit in a number of different ways. In this particular case, involving two different companies in the same industry:

- The responsible investor can use exit by selling the stock of both companies and making a strong public statement that these companies are not making sufficient progress on environmental matters, and hoping that adverse publicity will prompt both companies to adopt better practices

- The responsible investor can adopt voice by engaging with the company that is more inclined to dialogue and in the hopes that this will open up more productive communications with the other, and that this dialogue with both companies will ultimately influence others in their industry

- The responsible investor can use exit with one, hoping to enhance voice with the other – for example, by divesting from the company that has not entered into dialogue, thereby punishing it, and hoping that this will encourage the more cooperative firm to distinguish itself from its competitor by increasing the communications for which it has in effect been rewarded

Exit is more impersonal than engagement in the sense that it eliminates face-to-face contact between the shareholder and the company. Nonetheless, responsible investors who sell their stock can still indirectly pressure on a company by publicising their decision for doing so. For example, in June 2006, the Norwegian Government Pension Fund decided to divest from Walmart, primarily due to its dissatisfaction with the company's labour and human rights records.[67] The decision received widespread publicity and in turn led to further dialogue between the fund and the company.

Two conditions are necessary for the exit option to have practical effects. First, the company must be aware of the reasons for which its stock is being sold. Second, the sale must provoke sufficient public controversy to affect the company's reputation or to drive down its stock price.

The range of options for voice extends from the relatively quiet – a private letter, for example – to extended conversation, to highly public protest. The effective use of voice requires a nuanced articulation of one's critical opinions. However, voice can be time-consuming in both the short and long run, since meaningful change typically does not take place overnight.

In practice, exit and voice are not necessarily mutually exclusive. Indeed, they can gain in effectiveness if used in combination. Voice can alert the company to an issue that needs to be dealt with, while exit is held in the background as an option to be used if the company does not react appropriately. The timing of the use of these two options can therefore provide investors with a substantial array of choices.

Variable factors

Responses to this case may differ depending on a number of factors, such as the resources available for engagement, the engagement policy of the institution, the possibilities for maintaining some influence even when divesting and the severity of the concerns.

- What are the resources available for engagement?
  - What is the time-frame of the client or the money manager?
  - How much time and how many resources can the money manager devote to the engagement process?
  - How soon is improvement by the company expected?

- What is the general policy of the institution on engagement and divestment?
  - Is it the institution's practice only to engage with companies, never divest for societal or environmental reasons?
  - Can the institution act on behalf of all the investors for whom it is managing money?
  - Are press releases relating to engagement or divestment an acceptable practice for the institution?

- Is it possible for the institution to maintain limited influence after divestment?
  - Does the institution hold shares in the company in question through funds other than the RI fund?
  - Is the institution a member of collective initiatives through which it can exert influence on the company?
- What is the severity of the concerns?
  - Do the practices of the company breach negative criteria of the fund?
  - Are the concerns long-standing or recurring problems or are they isolated events?
  - What is the potential financial impact of the concerns?

Recommendations

As you approach dilemmas of this sort, you may want to:

- Determine the point of maximum leverage and influence with the company
- Be prepared either for a public confrontation with the company or for a less public engagement that can stretch over a long period of time
- Be clear and transparent about your engagement policies with your clients and with the companies involved
- Make appropriate use of both the exit and voice approaches as circumstance dictate

The voice and exit dynamic

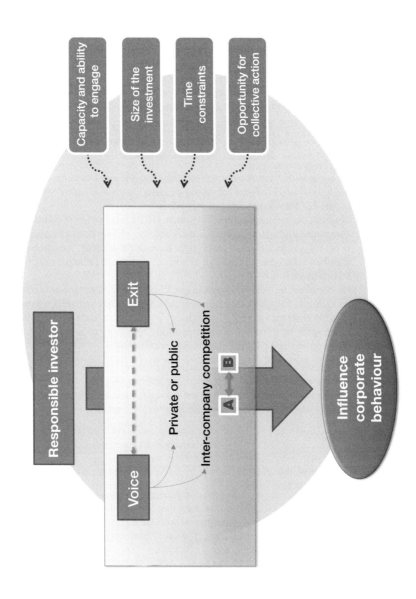

Responses from practitioners

Influence the company through dialogue

Dear Mr Thomas,

We have researched the environmental records of the two chemical companies that we hold in the separately managed account we maintain for your foundation. Although we believe that both these companies are addressing many of their environmental challenges, we understand your position that progress for both is frustratingly slow and can be improved on.

As you know, we regard efforts to influence companies to change as an important part of our work. To do so we often use the following tactics:

1. To begin with, we make our research assessments and scores for these companies publicly available. These are on our website. Companies know what they need to improve if they want to be in our portfolios.
2. Through our non-responsible investment (RI) funds, we are in contact with the companies that are not in our RI funds and can address corporate social responsibility (CSR) issues with them as an investor through these funds. We can potentially maintain a dialogue on CSR issues with all companies, whether or not they are in the RI portfolio.
3. On occasion, for companies not in our RI portfolio, it can take only minimal progress on their part to qualify them for inclusion. With these companies we make a particular effort to enter into a detailed dialogue to push them on their CSR efforts.
4. Often we write a letter to companies asking for clarification on specific issues to better understand their perspective and performance.

As you can see we do our best to influence companies' behaviour by informing them about our screening, and by giving them the reasons why they are not in our RI portfolios. Sometimes we issue press releases – for example, when a company is going to be added to our RI portfolio because of impressive improvements.

To respond concretely in this specific situation, we have considerable flexibility. This portfolio is customised to the specific concerns of your foundation and we of course want to be certain that you are satisfied that the environmental goals for your investments are being met. With this in mind, we would like you to consider two different approaches.

Approach 1: We could exclude both companies. Companies ought to be aligned with the criteria for your portfolio. If not, they should be excluded. It is as simple as that. But of course you will want to be transparent with the companies on the methodology and criteria that result in their exclusion. You may also want to be clear with the public on the reasons for which you have made this decision.

Approach 2: You may keep the stock of the company that is willing to talk because through dialogue we have some chance to improve its CSR performance. Dialogue can certainly be an effective tool. It has the potential to bring about change and to improve the behaviour of companies. But the results are never certain.

We would like to discuss in depth these two approaches with you. Would you please let me know of your availability in the coming weeks?

Sincerely,

Your Money Manager

Reaching the disclosure threshold

Dear Mr Thomas,

We appreciate your concern about the two chemical companies that are in our funds, your desire to see greater progress on both their parts when it comes to the environment and, in particular, the lack of transparency and willingness to communicate on the part of Company B.

As a responsible investor, we recognise the importance of transparency. We include transparency and willingness to enter into dialogue in our assessments and ratings of companies. If a company is a laggard in transparency in its sector, we may exclude it. In this particular case, we agree that Company B has shown itself to be a laggard.

We agree with you that Company B's unwillingness to enter into dialogue is cause for considerable concern. Before we exclude the company, however, we feel it necessary to take two steps. First we wish to attempt one more time to solicit a positive response from the company. However, you should understand that we will not be entering into dialogue with this firm explicitly on your behalf. Our policy for engagement is to always talk to companies on behalf of all our clients. We believe that engaging with companies on behalf of all our clients gives us a stronger voice.

You must also understand that if we feel the company's response is still not satisfactory, it will be because it has not reached a threshold for disclosure that is consistent with the overall criteria of our funds, not merely your particular level of concern. For consistency's sake we would be unwilling to sell our holdings in Company A, assuming it meets our threshold on transparency. We strongly believe that it is as a stock owner that we have our best chance of influencing companies on these issues. We are usually reluctant to sell its stock as a means of expressing dissatisfaction with the pace of progress.

If you have any questions, please do not hesitate to contact us. Before taking any action we will wait for your response.

Sincerely,

Your Money Manager

> 66 Many companies that we have engaged with for a long time have made very slow progress, but actually after several years, you realise that it was worth it. 99

Give the company time to change

Dear Mr Thomas,

Thank you for raising your concerns about the questionable environmental performance of two chemical companies that are in the portfolio of our fund in which your foundation is currently invested.

We will alert the companies to the fact that within the next six months to one year they must show concrete positive results. We understand that we need to give them time to change: first, to change policies and second, and more important, to achieve actual results. We are fully aware that progress doesn't happen overnight.

If we decide to sell a company out of our portfolio because it fails to communicate or fails to make adequate progress, please note that we would not make public announcement to that effect. As a money management company, we do not consider that as a part of our role. However, your foundation should feel free to comment publicly on the company and the appropriateness of our decision.

If you need further information or explanation, please do not hesitate to contact us.

Sincerely,

Your Money Manager

> 66 Exit is a last resort, because you lose the ability to influence. 99

> 66 Responsible investors have earned a certain amount of influence because as they walk out the door they can tell people about it. 99

In the news

Shareholder responses to companies concerning environmental, societal and governance (ESG) practices can depend on whether the company is transparent about, and willing to engage over, key societal and environmental issues with concerned shareholders. This willingness can play an important role for responsible investors in deciding whether or not to hold a company's shares. For companies that do not disclose ESG information or do not demonstrate a willingness to engage with investors, the only tool for investors to use for leverage may be to divest.

The following companies offer contrasting examples of approaches to CSR disclosure.

As of April 2010, JCPenney had several pages on its corporate website related to its CSR performance. In addition to the company's 2009 CSR report, the Social Responsibility homepage provided links to the company's CSR News Releases. Information about the company's performance was divided into four sections – Community, Associates (Employees), Responsible Sourcing and Environment – each of which had a page that provided details on the company's activities in those areas. The 2009 CSR report provided email and postal addresses for parties interested in more information about JCPenney's performance, indicating that responsible investors might well engage the company on a variety of CSR issues.[68]

As of the same date, Urban Outfitters, Inc., had no CSR content on its website. It did not publish a CSR report, provide a contact for investors interested in its ESG performance or make any mention of CSR. Responsible investors concerned about Urban Outfitters' ESG performance, particularly those investors with concerns about sweatshop or child labour in the garment industry, might have a difficult time entering into a dialogue with the firm.[69]

Cases for comparison

Compare and contrast this case with Case 6 'When a company changes', Case 7 'Public versus private partnerships for engagement', Case 9 'Incomplete societal and environmental data' and Case 10 'Exclusion of industries'.

Case 4

Societal returns versus financial returns

- **What constitute societal returns?**
- **What happens when there appears to be a trade-off between societal returns and financial returns?**
- **How can societal returns be measured and reported?**

The case

You are an RI fund manager at a major bank. One of your clients, Peter, is concerned about environmental issues. He wants you to move investments in his portfolio from airlines to railways, because railways are a more energy-efficient mode of transportation. At the current time, the prospects for financial returns from railways are less than from airlines (railway stocks haven't done as well over the past year and aren't projected to do well in the coming year).

You go back to Peter and explain you can make the change in the investment in the transportation sector of his portfolio from airlines to railways, but it may hurt the portfolio's short-term financial performance.

In order to make up his mind, Peter asks you whether he will be getting an equivalent 'societal' return from the investment in railways to compensate for the poorer financial performance.

You have a meeting scheduled with Peter. How are you going to answer his question about 'societal' returns . . . ?

Dilemma for the responsible investor

This case highlights the dilemmas that arise for responsible investors when, for a variety of different reasons, they need an accounting of the societal and environmental benefits that their investments have provided. Giving such an accounting can be a challenge. The financial community has developed sophisticated tools for reporting financial returns. However, little work has been done to date in establishing methodologies for reporting on and analysing the societal and environmental benefits of these same investments.[70]

The case in perspective

This case highlights what is often perceived as a dichotomy between financial and societal returns. Those who focus on price in the short term often forgo long-term perspectives, both financial and societal. This case illustrates how the consideration of societal and environmental concerns helps focus investment decisions on the long term, where the distinction between societal and financial benefit tends to be blurred.

The increasing short-termism of today's markets undervalues long-term thinking that can benefit corporate management, society and the environment. With a longer-term perspective tends to come an understanding of how financial and societal considerations coincide. What are often portrayed as contradictory goals become complementary ones.

The long-term perspective, however, increases uncertainties. Markets can easily make short-term mistakes about what is societally, as well as

→

Traditional money managers with a long-term investment horizon can encounter a similar situation when they identify an investment theme that they believe will pay off in the long term, but won't outperform in the short term. Unlike day-traders seeking to profit from short-term moves in and out of a stock, longer-term active managers may take a 'value' approach, basing their decisions on long-term cycles of certain industries or on the belief in the long-term ability of a company to outperform the market, despite recent periods of under-performance. In these cases, long-term investors justify periods of inferior financial returns because their financial models predict superior financial returns in the long run.

Responsible investors can find themselves in a similar situation in that they may hold stock in one company or industry that is under-performing financially but in which they believe in the long run. Their investment may be made on the grounds that the company or industry has a business model that the investor believes has a strongly positive societal or environmental story that will benefit society in the long run, although it may be difficult to quantify how these long-term benefits will translate into superior financial returns in the short run.

For example, given two apparel manufacturers, one using sweatshop labour and the other that makes a point of paying a living wage, a responsible investor might well choose the latter on the grounds that paying decent wages is societally beneficial in the long run, despite the fact that in the short run the costs savings the former achieves may give it a financial advantage. Similarly a responsible investor may believe that investing in alternative energy production – for example, solar power – will ultimately be profitable as well as help address the challenges of climate change, although it may financially under-perform coal, for example, for an indefinite period of time.

financially, beneficial in the long run. Responsible investors are forced to contend with these uncertainties as they attempt to distinguish long-term benefits from short-sighted profit seeking. These efforts remind investors that in the long run financial, societal and environmental considerations are inextricably bound together and that a market that disassociates them is likely to allocate its assets poorly.

The dilemma here is that, while traditional long-term investors make specific projections of long-term financial outperformance, responsible investors may not be able to quantify their societal and environmental concerns in financial terms. Projections of long-term outperformance may depend on such unpredictables as changes in government regulation, government subsidies, rates of natural resource depletion, the public's spending or consuming patterns, or perceptions of health, environmental or other cultural issues.

Responsible investors may not be able to place a specific current value on the benefits of their investment strategies. Yet they expect that in the long run these benefits will translate into value for society and ultimately for their portfolios.

Put more simply, responsible investors may well expect an accounting of the societal and environmental benefits created by their investments, although these returns are far more difficult to quantify, measure and report.

One complication in this calculus arises from the fact that societal returns take into account the interests of multiple stakeholders – employees, customers, suppliers, communities – in addition to investors. It is not immediately obvious how to calculate, cumulate and compare returns to these different stakeholders who may benefit in a variety of different ways.

RI can be viewed as a form of long-term investing that 'speculate[s] on the value of corporations to society and the environment, while simultaneously seeking to enhance that value at the company, industry and societal level'.[71] Responsible investors may be able to recognise the success of those speculations and quantify the results at a company level, but understanding how to achieve them at a societal level is less clear.

This case also raises a question as to whether the dichotomy between financial returns and societal benefits is in fact real. Current investment practice would have us express this as a trade-off between financial efficiency and ethical qualities of a portfolio. Perhaps, however, as Jed Emerson has put it: 'In truth, the core nature of investment and return is not a tradeoff between societal and financial interest but rather the pursuit of an embedded value proposition of both.'[72]

Approaches available to the responsible investor

One of the key themes raised by this case is the tension between the short term and the long term in investment and their relation to both financial returns and societal benefits. In confronting these issues, responsible investors can:

- Justify investment decisions where there are projections for short-term financial outperformance

- Relate investments to the possibilities for long-term financial outperformance that these societal and environmental benefits will ultimately generate

- Report on the societal or environmental benefits of particular investments independently from their potential for financial returns

- Measure and report on the success of an entire investment programme in relation to its ability to contribute to the realisation of societal and environmental benefits at the market or societal level

The first of these approaches avoids the complexities of societal and environmental returns separately. In effect, the responsible investor functions here much as does any traditional investor with the goal of beating the markets in the short term. For example, an investment in an organic baby food company that has an attractive environmental and health story might appear to be profitable in the near term. The stock is due to take off.

In the second approach, investments – for example, in the solar power industry – are part of an investment theme with a longer time-horizon, perhaps over several years. Like a traditional value investor, the responsible investor here may accept financial under-performance in the short term on the grounds that financial outperformance will ultimately materialise.

In the third scenario, responsible investors simply report societal and environmental benefits independent of stock returns. For example, the number of trees saved by recycling, kilowatt hours saved through energy efficiency, or lives saved through vaccine sales or donations. Financial value is not estimated.

In the fourth scenario, societal and environmental benefits are not quantified and measured in relation to a single investment or portfolio, but are conceived of as part of larger societal goals. Success is not defined in terms of cans of soup given away but in terms of reductions in hunger in a region. Only if investments provide benefits to society as well as to individuals can they properly speaking be justified.

The greater the weight given to societal and environmental factors and the broader the definition of success, the less distinct the line becomes between financial and non-financial returns and the longer the term in which benefits are recognised.

In addition, in considering the challenges of reporting on societal versus financial returns, it is helpful to distinguish between clients who put value (i.e. financial returns) first and those who place greater emphasis on values (i.e. societal and environmental returns). The greater the weight clients place on values, the more receptive they are likely to be to efforts to quantify the societal and environmental benefits of investment decisions.

Variable factors

Responses to this case may differ depending on a number of factors, such as the type of client, the materiality of the financial issues, how substantive the societal benefits are and the basic approach of the fund to societal and environmental issues.

- What type of investor is the client?
  - Is the client primarily seeking financial value or an expression of societal values?
  - Is the client an institutional or a retail investor?
- What are the potential financial trade-offs in the short term and the long term between the two sectors in question?
  - What is the likely financial upside or downside over the next 12 months for the two sectors?
  - What are the likely societal and environment upside and downside over the next five years for the two sectors?
- How substantive are the societal and environmental differences between the two choices?
  - How much better are railways than airlines from an environmental point of view?
  - Is there a downside to promoting railways? Is there a case for investing in a third alternative that would be better in the long run than either railways or airlines?
- What is the RI research methodology of the fund?
  - Has the fund historically taken a best-in-class approach?
  - Has the fund historically set a minimal basic societal or environmental threshold that companies or sectors must cross to qualify for investment consideration?

Recommendations

As you approach dilemmas of this sort, you may want to:

- Be aware of clients' financial reward and risk profiles
- Discuss with clients how societal returns can most appropriately be defined and measured
- Discuss with clients whether they feel that societal and environmental returns are as important to them as financial returns
- Be aware of the interplay between the short term and the long term when it comes to both financial and societal outcomes
- Encourage clients to keep in mind that the further out they look the more likely it is that societal and environmental benefits and financial returns will coincide

Balancing societal and financial returns

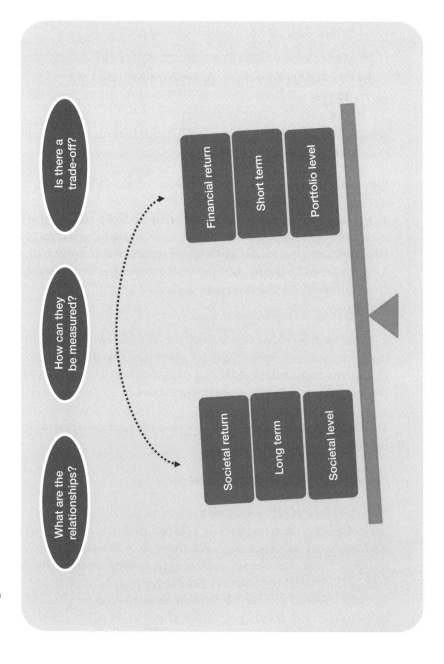

Responses from practitioners

Measure societal returns according to the client's approach

NOTES TO MYSELF IN PREPARATION FOR THE MEETING WITH PETER

What do I think Peter means when he talks about 'societal' returns?

Peter might define societal returns in one of two ways.

1. He might be interested in looking at the portfolio from a comparative standpoint: that is, he might want to take the comparative social ratings of individual companies and aggregate those into portfolio averages that he can then compare to the average social rating of another portfolio.

2. He might be thinking about the actual effect of the two industries as a whole. That is, what are the specific carbon emissions associated with the railways versus airlines? He could then construct a portfolio based on this comparison.

I would want to determine which of these two approaches he's thinking about.

What options should I offer Peter?

I could potentially construct a whole series of different portfolios that had different allocations drawn from all airlines and all railways and then vary the ratios between the two. Once I came up with a measure of societal return I could apply that to each of those portfolios and then I could compare societal returns to projected financial returns. Then I could prepare a chart that allows me to see the specific trade-offs of societal to financial returns and ask Peter: 'What trade-offs do you want to be making and where on this chart do you want to be?'

Framing the issue around risk

NOTES TO MYSELF IN PREPARATION FOR THE MEETING
WITH PETER

Ideally I would like to create a scoring or rating tool to
capture the main notions of sustainability, and be able to
give specific scores for each industry to Peter.

I should impress on Peter that it is important in this case
to think about more than just beating an industry-based
benchmark in terms of financial performance. There
are many other considerations that should go into a
measurement of returns.

In this case, it may be useful to frame the issue as being
about risk. Peter is essentially saying railways are more
energy efficient, so they're better prepared for weather and
energy shocks. Right now, or at a given moment in time,
that may not be valued in the marketplace. This is the type
of factor that tends to be valued in crises.

> I need to explain to Peter that this situation is not
> just about expected financial returns in the short
> run. It is also about how protected the companies
> and the industry are from a potential catastrophe.

What I'm also trying to ask is:

'Financial return over what time period?'

You may not expect railways to perform as well over a short
time-span, but over a longer time-horizon, they may excel.
It's reasonable to look at the long-term return potential
and at each industry's risk management. Those can be
legitimately considered financial criteria.

If we're looking for long-term return then I would tell him
to buy railways.

BUT: it's up to Peter to decide.

Making societal and financial trade-offs

NOTES TO MYSELF IN PREPARATION FOR THE MEETING WITH PETER

Make clear to Peter that societal returns can be quantified but it is not an easy task

There is such a thing as a societal return, but it's incredibly hard to quantify. If you try to quantify it too precisely, you will run into difficulties. Sustainability is an investment factor that is difficult to quantify. You can use all kinds of regression analysis to try to parse a portfolio that is run according to sustainability relative to a mainstream portfolio, but generally speaking the sustainability portfolio has multiple factors driving its performance. These could be growth or could be linked more to its exposure to one region or another. So it's hard to pinpoint what exactly is attributable to societal or environmental factors and what is not - but it can be done.

My approach/suggestion

I suggest that Peter look primarily at the environmental perspective of shipping via rail versus airlines. There are plenty of databases we have access to for this data. I can look up the GHG [greenhouse gas] emissions of each sector and the companies within the sector and report carbon emissions to show that the environmental impact of rail versus airlines is much less. I can quantify the extent of the difference of that impact, and I can also try to monetise that difference relative to current GHG emission costs. I can also use historical measures of fuel usage and historical environmental records to look backwards and show the potential environmental liability of airlines relative to rail.

page 1

In this case, Peter will have to make <u>societal and financial trade-offs</u>.

 Is he comfortable making that bet?

If Yes, I can provide metrics that show financial valuation versus environmental performance. I can show that airlines look very attractive in the near term, but have a long-term liability given GHG regulations - which we can quantify.

Given the poor business model of airlines and impending environmental legislation, in the near term airlines may look good but rail is much more compelling long term. In the long run, I should recommend moving to rail.

> ❝ Rather than separating societal and financial issues, we need to look at them in an integrated way over the long term. ❞

> ❝ A lot of people discuss this, and some say there is a trade-off, but we don't believe there is over the long term. Over the long term, these issues will be reflected in the financial value, and that's how we define sustainability issues. ❞

> ❝ Since we cannot calculate the environmental and societal impacts precisely, we would make some qualitative statements of the pros and cons and formulate the main arguments that can be used when we try to balance the societal and financial implications of our decision. ❞

In the news

When it comes to the stocks of individual companies or of industry sectors, there are often periods of time when the stock performance of a company or industry with the poorer societal or environmental profile surpasses that of a company or industry with a superior profile. These periods of time can be substantial. For example, from 5 May 2009 to 5 May 2010 the Dow Jones US Railroads Index, which included companies such as CSX, Union Pacific Corp. and Kansas City Southern, rose approximately 60%, while the US Automobile Index, which included companies such as Ford, Harley-Davidson, Honda and Volkswagen, rose approximately 90%.[73] RI investors concerned about environmental impacts within the transportation sector might have chosen to invest in the railway industry, which is notably energy-efficient when it comes to transporting goods over long distance, and avoided the automobile sector, which at that time had been slow to address fuel efficiency and alternative energy issues in a systematic fashion.

Similarly, during that same period, the Dow Jones US Soft Drinks Index, which included companies such as Coca-Cola and PepsiCo, rose approximately 28%, while the US Water Index, which included companies such as California Water Services and America Water Works, rose approximately 5%.[74] Responsible investors have historically raised questions about the health benefits of highly sugared soda and many in the environmental movement have advocated drinking tap water, rather than bottled water or other less healthy beverages.[75]

Over this one-year period, therefore, responsible investors pursuing an investment theme that favoured energy efficiency in the transportation industry or health and environmental issues in the beverage industry would not have been rewarded in the stock market.

Cases for comparison

Compare and contrast this case and Case 10 'Exclusion of industries or companies'.

Case 5

Alleged versus confirmed illegal activity

- **When should responsible investors take a stand on illegal company activity?**

- **How do responsible investors know when a company has acted illegally?**

- **Are there degrees of seriousness between corporate illegal activities?**

The case

You are a money manager with a client, Ms Harris, who is concerned about having only companies that are run ethically in her portfolio. Recently two companies in her portfolio have appeared on the front page of the newspapers. Company One is accused by governmental regulators of massive anti-trust violations – unfairly driving competitors out of business and then raising prices on its products. This is one of the largest anti-trust cases in recent history, but the company denies all allegations, asserting that it uses totally fair tactics to win in the marketplace. This case is likely to drag on for several years before it is resolved. Company Two has just settled a massive bribery case in the course of which its CEO and several top executives resigned. The company has simultaneously announced that a comprehensive ethics and anti-bribery system is now being put in place throughout the firm. The case is over and settled.

Would you sell one or the other of the stocks? Both? Or neither . . . ?

Dilemma for the responsible investor

This case highlights the dilemmas that arise for responsible investors when they seek to take illegal corporate actions into account. Responsible investors may wish to assess the financial, ethical, or societal implications of illegal behaviour by corporations, but determining if a company has broken the law, and how serious its infractions are, is not always simple.

This is similar to situations that traditional money managers encounter when they are concerned about the effects of pending litigation on companies' financial viability and stock price. For example, when a drug company faces class action lawsuits regarding the safety of one of its products, investors may sell the stock if indications are that lawsuits will go against the firm. Similarly, the stock of Philip Morris (now Altria) had a rare decade of underperformance from 1992 to 2002 when it appeared that tobacco litigation might result in substantial legal claims against it.

The mainstream financial community may often be concerned about the financial outcome of litigation, but it is generally not interested in how lawsuits can be used as a window through which insights can be gained about a firm's corporate culture or the ethical quality of its management. From the financial community's point of view, it is the job of government to identify illegal activities, prosecute them and assess appropriate financial penalties. The mainstream's primary interest lies in whether, and to what degree, prosecution for illegal activity will affect a company's finances or stock price.

In addressing the question of corporate social responsibility, Milton Friedman famously commented that 'there is one and only one social responsibility of

The case in perspective

This case focuses on responsible investors' assessments of the interactions of legal and regulatory systems with corporations. It acknowledges the need for these investors to exercise judgement when it comes to the law and legislation and, simultaneously, the fact that they cannot always look to a specific outcome of a particular case to understand whether corporations are or are not playing by the 'rules of the game'.

Corporate confrontations with the law are a cause for concern. Their financial implications are important to all investors. They can also provide responsible investors with valuable insights into the culture of the corporation, the quality of management and the direction of the firm. Although some cases are clear-cut and relatively easy to assess, others involve complicated factual situations, claims and counter-claims, behind-the-doors deals and sources of ambiguity that make them a challenge to evaluate.

> These cases remind investors that their assessments of corporations' relationship to society and the environment necessarily take place in the context of complicated legal and legislative frameworks and shifting political landscapes. Their judgements about corporations cannot be separated from their assessments of these frameworks. These assessments of the legal structures within which corporations operate are part of the informed debate about the proper role of corporations in society in which responsible investors take an ongoing part.

business – to use its resources and engage in activities designed to increase its profits so long as it stays within the rules of the game, which is to say, engages in open and free competition without deception or fraud.'[76]

Responsible investors wanting to understand if a company has played by 'the rules of the game' face a myriad of dilemmas. To begin with, corporations are accused of breaking the law all the time – sometimes fairly, sometimes unfairly, sometimes by government, sometimes by consumers, sometimes by their competitors. It can become clear that they have in fact broken the law when the courts rule that they have done so, but companies may also agree to pay a penalty and settle a lawsuit without admitting guilt. Should they be penalised for that? Companies may be convicted by one court but appeal the case to a higher court. Should the responsible investor act on the first court's ruling without waiting for the higher court to rule? A charge may be brought against a company but the case may not go to trial for years. Should a responsible investor wait years before acting while the case makes its way through the court system?

By the time a company is convicted by the courts, the executives responsible for the illegal activities may be long gone and the company may have initiated thorough programmes to correct its abuses. Should a responsible investor take action so long after the original charges and after so much has changed? A company may be accused, or convicted, of illegal activities, but all the major companies in its industry were engaged in the same illegal practices at the same time. Should the responsible investor single out only one company – for example, the worst of the violators – or act with regard to a whole industry? Big companies tend to pay big fines when convicted, and smaller companies smaller fines. Should a responsible investor attempt to assess the seriousness of illegal actions by the size of the fines or penalties?

Moreover, responsible investors recognise that at any given time government may not have established laws or regulations to address various problematic aspects of corporate behaviour. In addition, industry may lobby successfully to weaken, derail or postpone enactment of legislation that would regulate some of its questionable activities. Everything that is currently legal is not necessarily everything that is ethical.

In short, relying solely on decisions by government, regulators and the courts to determine who has, and who has not, played by the rules of the game is an imperfect tool.

If responsible investors are to ask that companies act within the spirit, not just the letter, of the law, they must use their judgement in many cases. Judgement entails personal judgement here, and there are no firm rules for applying it.

Approaches available to the responsible investor

One of the key themes raised in this case is that of judgement. Because of the complexities of the legal systems, responsible investors must often exercise judgement in assessing allegations of illegal behaviour. Among the approaches that a responsible investor can take to accommodate the simultaneous demands of judgement and objectivity in these situations are the following:

- Have a consistent policy of how to approach legal cases

- Analyse the legal implications oneself

- Consult outside experts and impartial third parties

- Discount or ignore certain types of lawsuits

As appealing as it may be to rely on the considered judgement of the courts when assessing the appropriateness of corporate actions, it is often necessary to make decisions before the law has completed its due process. In addition, even when cases have been put to rest, the ultimate guilt or responsibility of corporations for alleged crimes or violations is not always clear because many cases are settled out of court without any admission of guilt.

The demands for consistency and objectivity make it tempting for RI analysts to adopt a clear and consistent policy for legal cases – for example, 'we will always wait for the court's final ruling'; or, conversely, 'we will always make our judgement based on the best information available at any given time'. In practice, however, neither approach is entirely satisfactory. The first can lead to withholding judgement for years while massive, high-profile lawsuits work their way through the courts. The second can prompt decisions that may need to be revised when new facts emerge or when the courts eventually reach their verdicts.

It is usually a good policy for RI analysts to research the details of allegations. Such research deepens one's understanding of a case and helps in assessing the legitimacy of both sides' positions. Nevertheless, this research can be time-consuming and by no means guarantees a clear-cut understanding of the relative merits of the case.

It is also helpful to consult outside experts or other impartial third parties on their views about a case. This is a variation on the approach of analysing the allegations oneself. Finding trustworthy third parties can also be a challenge.

Finally, it is possible to distinguish between different types of lawsuit and court battle and to consider some more serious than others. For example, lawsuits between competitors over patents, intellectual property and anti-competitive business practices frequently occur. Similarly, in the United States, cases are frequently brought against publicly traded firms when their stocks falls dramatically in response to an unfavourable earnings report, with allegations of prior inadequate disclosure by management. These two types of lawsuit may be regarded by responsible investors as a normal part of doing business and, unless exceptional in size or scope, may be considered less informative about a company's ethical qualities and broader relationship to society than a lawsuit alleging such things as bribery, severe environmental harm or poor product quality control that results in serious harm to customers.

These various approaches, used alone or in combination, can provide a much-needed flexibility in making judgements in these often difficult cases.

Variable factors

Responses to this case may differ depending on a number of factors, such as the jurisdiction within which the case is being brought, the frequency of such cases among industry peers, who is bringing the case, whether the company has faced similar cases in the past, and how the company has responded to the accusations.

- What is the legal context of the jurisdiction or industry within which the case is taking place?
  - Is this type of lawsuit frequent in the region?
  - Are the proposed or assessed fines large or small relative to other fines in similar cases in the region?
  - Are the alleged misdeeds typical of the company's peers as well?
- Who is bringing the case?
  - Is it a competitor of the company?
  - Is it a single individual or has the case been certified for class action status?
  - Is it a regulator or a governmental body?
- Are the allegations part of a pattern or a one-time event?
  - Has the company been accused or convicted of the same misbehaviour before?
  - Does the company have a pattern of being the subject of lawsuits across a number of societal and environmental areas?
- How has the company responded to the allegations?
  - Has it fought the charges or agreed to address the allegations?
  - Has the company taken steps to correct the alleged problems?

Recommendations

As you approach dilemmas of this sort, you may want to:

- Be aware that legal cases can have many ambiguous aspects
- Understand the specifics of the legal systems and norms of the countries within which a company is operating, because the similar situations can be treated differently in different nations or regions
- Assess the relative performance of one company relative to other companies in the same industry

- Distinguish between lawsuits that arise between companies over business practices (less serious) and those brought by governments on issues of broader public interest (more serious)

- Look for patterns of illegal behaviour and distinguish them from a single legal action against a firm

- Evaluate the company's response to the allegations, assessing the legitimacy of its claims and any indications of reforms it has undertaken

Decision tree for illegal corporate actions

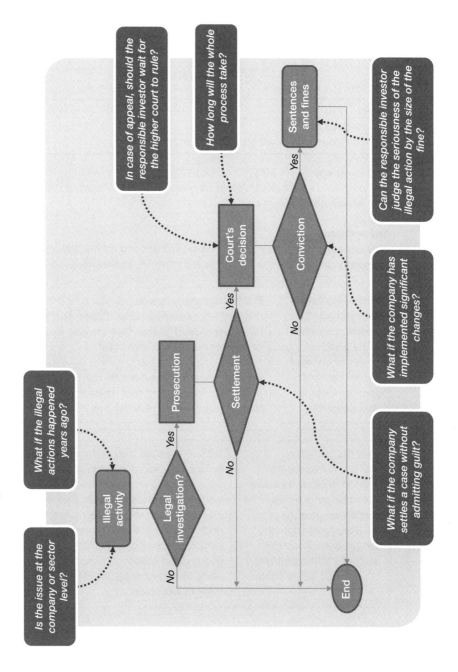

Responses from practitioners

The importance of the company's history

Memorandum

From: Bill
To: Sarah
Re: Your enquiry re: Ms Harris' Portfolio

2nd company
The company's behaviour should be rated positively. We would certainly not exclude the company.

1st company
This case is more difficult. Our decision will depend on the **history of the company** and whether or not there are previous, similar allegations.

| **A** If the company has not had similar problems in the past, we need not conduct further research and can include the company. | **B** If the company has had similar problems in the past – for example, other cases where they have been found guilty – they may be excluded or at least put on the 'watch list'.

↓

We also need to consider how they dealt with this particular problem and whether they have tried to put the right new policies in place. |

| **B1** If the company previously had weak policies in place, but now has stronger policies, we can assume they are likely to change and can probably keep the company in the portfolio, other things being equal. | **B2** If in the past they have put new policies or programmes in place and no real change has occurred, then we face a difficult decision. One thing we might do is look at the rest of the sector. Is this problem common to the sector, or specific to the company? If this problem does not arise for other companies in the sector, we should think of exclusion. |

- page 1 of 2 -

Key questions:

Q1. What is the history of the company?
It depends on: ⟶

- Number of allegations
- Size of the fines or penalties
- In one country or everywhere?
- Subsidiary problem or at headquarters only?
- If due to a subsidiary/branch they have just bought?

Q2. Would we wait for the end of the trial?
It depends on the past history of the company. When a company is placed on the watch list it is theoretically for several months. This time is mainly used for us to conduct further research investigations. It is difficult to wait for the end of the legal process as it can take several years.

Q3. Can we still trust the management? What will the financial impact be?
A settlement is good but it leaves much uncertainty and in a way makes it more difficult for us to judge what has happened. It may well be possible that the problem remains.

Your assessment of this situation for Company One may not necessarily lead to its exclusion, but it will certainly be one of the factors that we will take into account.

> 66 I think it comes down to the quality of management currently in place at a company. It becomes very much a judgement call. But it ultimately comes down to whether they have the right people to run the company in an ethical manner. 99

> 66 I would look for patterns of allegations, as well as evaluating this one particular situation. If this is the only allegation, I would treat it differently than if there is a pattern, if the company has had several allegations. 99

Patience rather than rushing in

Memorandum

From: Carla
To: Tom
Re: Your enquiry re: Ms Harris' Portfolio

I'm glad to share a few thoughts with you on this situation. It is a difficult one that arises frequently and I have some preliminary thoughts, as well as some further questions.

→ <u>Hold on to both companies</u>

- For the first company, the allegations are just being made and are unresolved. We would want to know on a financial as well as a social basis: is this about the corporate culture, or just a bad apple or two? You can't hold the company responsible for an individual's bad choice unless the company is feeding that somehow.
- The second one we would hold because if changes have been made and if they're real, then I don't see it as a problem. We want to explore in more depth whether the changes are being made in a timely fashion and whether things are going the way they ought to.

If this were a co-mingled product, then there would be an investment policy statement with an overarching policy in place and our approach as a management company would be relatively clear. We would have an obligation to stick to our stated values and responsible investment philosophy. We would still need to be cautious in our approach, though. We tend to take a wait-and-see attitude and don't always trust media reports in these cases.

However, because this is a client with a separately managed account, we will have to look a little closer. The client came to us for the kind of customised account that we as bank trust officers can offer individuals wanting help in integrating social responsibility into their investments. We are in a different position from a mutual fund that has a single set of societal and environmental guidelines that it is publicly committed to implement consistently. Because we are working within a large firm with no corporate guidance on this particular issue, we can tailor our decisions to the individual client's values.

There is nothing easy about these kinds of questions. When do you decide to hold a company that has gone down in flames, but then has come back in a new and substantially changed form with new management? Generally speaking, because we are patient, long-term investors, we tend to hold companies in complicated situations such as these and seek to understand the situation, rather than rushing to assume that there is an overall failure of ethical culture inside the company. However, each case is different and needs to be evaluated on its own merits.

In the news

The following examples demonstrate contrasting responses of two large multinational corporations to major lawsuits alleging illegal behaviour.

Microsoft's ten-year battle with the United States and Europe over anti-trust allegations was frequently front-page news. The first suit, a two-and-a-half-year battle with the US Department of Justice and 20 states, was filed against Microsoft in May 1998. The government's anti-trust suit alleged that Microsoft abused its power of monopoly as the only operating system on Intel machines by including its Internet Explorer browser with its operating system. This bundling effectively gave Internet Explorer control in the browser market, as competing browsers, such as Netscape, faced prohibitively high barriers to reach consumers. Microsoft denied any wrongdoing and fought the suit. A US District Court judge found in favour of the plaintiffs in 2000, but Microsoft appealed and the ruling was later partially overturned. The parties finally reached a settlement in November 2001.

Microsoft's legal imbroglios did not end there. The company continued to be named in lawsuits, notably in the European Union. Microsoft's battle with the European Commission resulted in a protracted ten-year fight over the bundling of Windows Media Player into the Windows operating system. The suit and other disagreements stemming from the court's ruling finally ended in 2008, with Microsoft paying nearly €1.7 billion in cumulative penalties. No top Microsoft executives resigned as a result of these lawsuits.[77]

Siemens, faced with similarly serious allegations of impropriety in 2008, took a different approach. *U.S. v. Siemens AG* and *U.S. Securities and Exchange Commission v. Siemens AG* were the US suits brought against Siemens for allegedly paying bribes and kickbacks to foreign officials in 12 different countries, beginning in the 1990s, in exchange for government contracts. Siemens opted to settle within days of the filings, paying $800 million in fines to US authorities. In Europe, an additional $800 million was paid to German authorities, which agreed to terminate legal proceedings against the company.[78]

In the aftermath of the bribery scandal, Siemens' CEO stepped down, half of the company's top 100 executives were fired and the company was restructured to prevent future misconduct. By settling, the company sought to avoid potentially higher fines, reputational risks and restrictions on government contracts that might have occurred if they had fought the case.

Cases for comparison

Compare and contrast this case with Case 6 'When a company changes', Case 11 'Emerging issues' and Case 12 'Privatisation of public services'.

Case 6

When a company changes

- What happens when a company with a reputation for good or bad behaviour appears to change?

- How can one identify meaningful societal or environmental change in a company's record?

- When should responsible investors add or remove a company from their portfolios in response to such changes?

The case

You are the manager of an RI fund. Ten years ago you sold out of your portfolio a large, successful retail company that was criticised for a series of abuses: harming local communities by driving small businesses into bankruptcy; using low-cost suppliers in China that had abusive labour conditions; disregard for environmental issues; discrimination against women and minorities in the workplace; fierce anti-union policies; and so on. For years the company showed no interest in corporate social responsibility (CSR) and became the subject of a worldwide anti-business campaign. However, in the past two years, the company's leadership has decided to make CSR a top priority. It has become a leader on sustainability (energy efficiency, organic foods and sustainable fisheries). It has imposed strict labour standards on its vendors in China and stopped fighting unions. It has appointed a woman CEO. Environmental and other activist groups still criticise this company and say it hasn't gone far enough for a company of its size and influence. Company representatives ask to meet with you to discuss its progress and the possibility that you might add it back into your portfolio

What do you tell the company . . . ?

Dilemma for the responsible investor

This case highlights the dilemmas that arise when responsible investors make judgements about whether a company's societal and environmental records have changed substantially for the better or the worse. Indications of change pose challenges for responsible investors both in determining how meaningful the changes are in actuality and in deciding at what point it is appropriate to take action in response to these changes.

Mainstream money managers frequently confront the question of corporate turnarounds from a financial point of view. If a previously poorly run company now appears to have better management or to have adopted a new, well-conceived business strategy, at what point does an active manager purchase the stock, betting that the company's financials will improve? Or conversely, if a well-run company appears to be veering off course, at what point will the manager sell its stock, foreseeing a deterioration in its profitability?

The challenge is similar for responsible investors evaluating a company with a previously poor or strong societal and environmental record that appears to change. At what point is it possible to say that a company's practices have meaningfully improved or deteriorated?

A relatively simple variation on this theme arises when a company has been involved in a single, clear-cut area of controversy – for example, it is doing business in one specific country with a record of extensive human rights violations; it has handled poorly a strike at one particular plant; or it is asked to phase out use of a toxic chemical in one of its product lines. The question of change is straightforward – does a proposed new course of action adequately address the concern? There still remain, of course, questions of timing.

The case in perspective

This case draws attention to the fact that evaluating changes in corporate culture is a challenging, but potentially rewarding, task that involves fundamental uncertainties. Forming reasonable judgements in these cases requires time and careful consideration. By contrast, the tendency of today's stock markets to focus on short-term financial results as indications of change in the value of corporations often leads to snap judgements and unreasonable oscillations in the company's stock price.

The value of a corporate culture involves intangibles that are difficult to measure and trends whose outcomes are often uncertain. Moreover, companies do not always make these evaluations easy for outsiders, presenting only their best face to the public, anticipating changes that have yet to occur, or hiding bad news. Even when the facts are clear, it can still be hard to know whether trends will continue.

The virtue of contending with these ambiguities lies in the incorporation of a measured and careful consideration of the underlying value and values of a corporation and its overall direction of change. Implicit in this process are relative stability in market judgements and orderly, incremental changes in valuations.

Should a responsible investor recognise the new policy when it is announced? When it has begun to be implemented? When it is completely implemented? When its results have been proven over time? Despite these challenges with timing, the specific changes desired can be measured and evaluated.

The situation is more complex when a company has been involved in a pattern of questionable behaviour. It can then be unclear whether proposed changes in a single area are primarily symbolic or substantive. Does one change indicate an overall turnaround is underway or is it just for show? If changes on one issue are being implemented systematically throughout the firm, that is certainly a good sign, but what does it mean if only one of multiple issues has been addressed? Or even if all issues are being addressed, how does one know if the company is committed to these changes for the long term? Conversely, for a company with a previously strong CSR record, is a single negative incident a sign of change for the worse or just an isolated incident?

In addition, the question of the timing in anticipation of such changes can be complicated. A mainstream money manager may decide a stock is undervalued, buy it while it is cheap and be content to wait patiently for substantive change to result in improved financials. Responsible investors may not want to act so soon, because they may not want to publicly reward a company for promised changes before they have been fully implemented. On the other hand, failing to reward company management for progress in the right direction might send the wrong message – corporate managers might feel that no matter how many positive initiatives they undertake, they will not be able to shake off their poor reputation, that responsible investors will never understand and appreciate their new efforts and that expensive initiatives to do the right thing will not be rewarded in the marketplace.

Contending with the public's perception of a company's past societal and environmental record can also pose challenges for responsible investors. A company that has long been controversial can have a poor or positive reputation firmly established in the public's mind. If responsible investors decide that the company has truly changed, they must manage their own stakeholders' understanding carefully, making a clear and convincing case for any new assessment of the firm's record.

Approaches available to the responsible investor

The key theme raised in this case is that of the assessment of the meaningfulness of apparent change and the timing of actions in response to that change. Changes in corporate policies, practices and even cultures can take place rapidly and clearly or slowly and in more ambiguous stages. Determining when such change is real and deciding when it should be acknowledged – either positively or negatively – is often a difficult task.

In assessing the meaningfulness of changes at a corporation and how they are perceived by the full range of stakeholders in the corporation, responsible investors can:

- Communicate closely with the company
- Communicate closely with stakeholders in the firm
- Set specific milestones that must be met to trigger action
- Look for more general changes in corporate culture

On the question of timing, responsible investors can:

- Act when changes first occur but their full significance may not be apparent, or
- Wait until the full significance of the changes is clear

Responsible investors can usually benefit from keeping in close touch both with companies whose record appears to be changing and with their stakeholders who have followed closely changes at the firm, for better or worse.

Communications will help responsible investors understand the nuances of the changes that are taking place and help in assessing whether these changes are systematic or simply anecdotal, and whether top management is altering the nature of the company's culture in one direction or another. The perception of stakeholders directly affected by the changes can be invaluable in these regards. If, for example, environmentalists or union leaders previously critical of a firm indicate that they believe recent changes are real and long lasting, that can be a particularly meaningful indicator.

In general, responsible investors want to minimise the subjectivity of their assessments of change. To help do so, they can set specific targets that a company must achieve before they will act – for example, meeting a specified goal for reductions in emissions of toxic chemicals or for completion of third-party audits of labour conditions at overseas vendors' plants. Once this goal is met, they then have a clear justification for their new assessment that they can point to, even if the public's perception of the company may not yet have changed.

In addition to these specifics, responsible investors may also want to wait for confirmation of a change in the corporation's culture. They may believe that progress on one issue can easily be undone by new controversies in another area if the root cause of a firm's problems – for example, a cost-cutting culture that cuts corners on safety – has not been addressed. They will then look for more general change in multiple spheres or for a new tone from the top. Assessments of this sort involve a substantial degree of subjective judgement and a sense or feel for the firm.

On the question of timing of actions based on assessments of change, by acting sooner rather than later responsible investors achieve a number of goals. If the changes are to the positive, they can demonstrate to corporate management that their efforts are recognised and appreciated. They can create momentum for new directions an industry may be taking, encouraging others to join in ground-breaking initiatives. They can strengthen ties between investors and the firm, build trust and create the potential for further change. If the changes are to the negative, responsible investors can avoid the risk of future bad news for the firm that may not be recognised by others.

Waiting for full implementation of new positive programmes, by contrast, helps ensure that companies don't abandon new efforts halfway down the road, minimises the chances of responsible investors being fooled by 'greenwashing' and public relations, and demonstrates respect for the concerns of those corporate critics who often may be reluctant to change their stance on a firm until they see substantive changes in fact.

This case also raises indirectly an additional theme – that of responsible investors serving as consultants or quasi-consultants to corporations. When dealing with societal or environmental challenges, corporate management may approach responsible investors for advice. What are the most important issues it should be tackling and how? Which options would be perceived as most meaningful by responsible investors? By what criteria is their company judged by responsible investors and what can they do to improve their 'score'?

It is not always clear how far an RI money manager, researcher or analyst should go in providing guidance to a company in these cases. A danger of loss of independence, or a perception of such a loss, exists if responsible investors become too closely associated with a firm. Nevertheless, responsible investors will want to take advantage of the opportunities to influence firms' behaviour that these situations provide. Ongoing dialogue can be an important catalyst for change, but responsible investors also need to maintain their independence and the credibility of their judgements about a company's record.

Variable factors

Responses to this case may differ depending on a number of factors, such as past and present allegations against the company, the type of changes that are required, the general attitude of the company, the fund's past approach to such situations and its clients' expectations.

- What allegations are we talking about?
  - Do they suggest massive violations or only a minor issue?
  - Is there a series of concerns or just one single concern?

- What type of change is required?
  - Does the issue require an incremental or transformative change?
  - Does it require an organisational, structural, technical or cultural change?
  - Does it relate to the company's own operations or does it involve their suppliers?

- How trustworthy are the company's positive changes?
  - Has the company addressed the concerns thoroughly?
  - Are the changes reflected only at the policy level or also in actual implementation?
  - Has the company communicated transparently about its new approach?
  - How committed are the company's top managers to the changes?
  - Are the changes implemented throughout the organisation?

- What is the client's attitude towards the change?
    - Is it the client's perception that the company has changed?
    - If not, how can the RI money manager convince the client of the reality of the change?

Recommendations

As you approach dilemmas of this sort, you may want to:

- Evaluate carefully the facts available from the company and the assessments of these facts by various stakeholders in the corporation

- Keep an open mind on the possibility that a firm can genuinely change for the better or the worse, despite firmly entrenched perceptions by the public one way or the other

- Be clear and consistent about what indications of change should be considered most meaningful

- Recognise that judgements made in these situations may turn out to be wrong and be prepared to change your decision in the light of new information

- Communicate clearly and thoroughly with all parties on the reasoning behind any actions taken in these cases

Capturing change

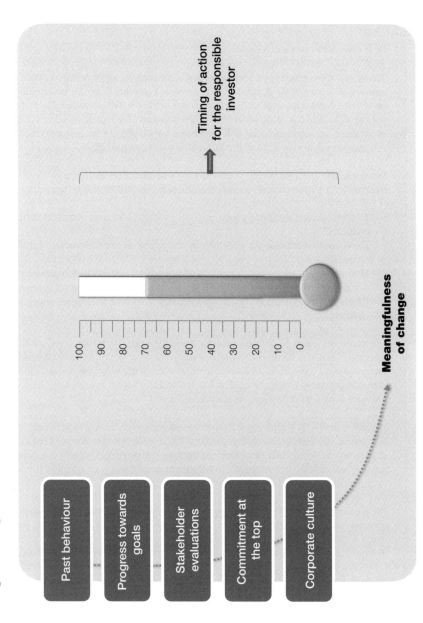

Past behaviour

Progress towards goals

Stakeholder evaluations

Commitment at the top

Corporate culture

Timing of action for the responsible investor

Meaningfulness of change

100
90
80
70
60
50
40
30
20
10
0

Responses from practitioners

Need for an objective assessment

Dear Company representative,

Thank you for contacting us and requesting a meeting for next month. As you know, a number of years ago we decided to exclude your company from our investment universe because of serious societal and environmental concerns. We are pleased to learn that you believe your firm has made significant improvements. We look forward to learning more about these improvements and will reconsider your company for potential inclusion in our investment universe in light of the information you can provide.

Since we have been following your company over the years and are aware of its historical track record on societal and environmental issues, we will be asking the following:

*   In what specific ways has your company's societal and environmental performance changed?
*   How has your company's reporting on its societal and environmental initiatives changed?
*   How does your company currently perform on key performance indicators in comparison with your peers?
*   How substantial is the company's improvement relative to its most important challenges?

Our goal is to assess as objectively as possible the actual performance of your firm. Your company's policies themselves, while praiseworthy, are not of primary importance to us. We are instead focused on what your company has actually accomplished, the extent of its actual improvements and the goals it has realistically set for itself. Please be prepared to provide us with hard data and facts in these areas.

Because our clients are invested in customised, separately managed accounts, we will summarise our findings on changes at your firm for them and make our final decisions in part depending on their perception of the data you provide. Our role is essentially to provide information to our clients, to help them understand your actual performance and to guide them to a decision with which they are comfortable.

We are confident that if the numbers that you can provide us with look good, if your company shows continuous improvements on its societal and environmental performance and if we can assess your performance relative to others in your sector, our clients will be in a good position to conduct a fully informed review and make a fully informed decision.

We look forward to our meeting with you and will let you know the outcomes of this review process.

Sincerely,

RI Analyst

Genuine change and lasting improvement take years

Memorandum

From: Karen
To: Ben
Re: Company X meeting and review

The following steps need to be taken for a reconsideration of Company X for inclusion in our investment universe:

Meeting the company
The objective of the meeting should be twofold:

1. Evaluate progress made by the firm beyond simple cosmetic changes. For that we need data, both current and historical, so that we can be sure they really are on a path to continuous improvement. Use our standard survey on key societal and environmental performance indicators for companies in its industry.
2. Make a subjective assessment of the integrity and effectiveness of top management at the firm. How committed are they to real and lasting change? Is this just a short-term improvement or can we be assured that the improvements will become part of the culture at the firm?

Benchmark the company against its peers
We need to assess the company's current performance relative to their peers. If they are better than their peers in all regards, then we could reach a conclusion more quickly than if they are average or still in a catch-up mode.

Verify our clients' expectations
How well do we understand our clients' expectations with regards to this firm? Our clients come to responsible investing with different mind-sets and differing expectations. It is important that we know our clients well, and have a feel for how independent they are in their judgements, how much they trust our judgement, and how influenced they are by the media and others who have historically been critical of the company.

- If this were an institutional client, a university pension fund, for example, we would have met with them regularly; we would know the people; and we would have debated similar investment cases over the

- page 1 of 2 -

years. We would then have a clear basis for our understanding of the concerns as we make our decision.
- Because this is a retail fund, however, with many investors who come with all sorts of prejudices and opinions, we must approach our decision-making and communications differently. We must be able to understand and articulate in general their past objections to the company, be sure that the changes at the firm address these objections and then communicate clearly why we believe the changes are real and long-lasting.

In short, to include the company in the portfolio we need to be convinced that they have genuinely changed and that it is a genuine reflection of a different attitude of the business. Please keep in mind that these kinds of lasting improvement often take years before they become a part of the company's corporate culture.

> 66 We need to evaluate progress made by the firm beyond simple cosmetic changes. For that we need data, both current and historical, so that we can be sure they really are on a path to continuous improvement. 99

> 66 Basically companies should be expected to be able to make substantial strategy changes. 99

> 66 If the company is creating value or starting to create value through being more efficient around environmental issues, for example, there is definitively a value creation. 99

In the news

Walmart is one of the world's largest and most successful businesses and one of the great American corporate success stories. In its nearly 50-year history, Walmart has gone from one privately owned store in Arkansas to a publicly traded international chain of over 6,200 discount superstores.

As one of the most visible corporations in America, Walmart has also been subject to some of the most bitter criticism of corporate behaviour. In 2005, labour unions, environmentalists and community groups joined together in a concerted, coordinated anti-Walmart protest. The Walmart campaign alleged poor labour relations, suppliers' human rights violations and negative impacts on the environment and community relations, among other things, through a series of documentaries, publications, shareholder resolutions, boycotts and lawsuits – all of which received wide press coverage. The company rapidly came to be perceived as a poster child for corporate irresponsibility.

Recognising the reputational and financial risks of ignoring this coordinated campaign, Walmart set about systematically countering this campaign. In addition to vigorous public relations efforts, its chief executive officer declared corporate social responsibility a top priority at the company and launched a number of initiatives, including a programme called 'Sustainability 360' aimed at helping the company become more environmentally sustainable via renewable energy, reducing waste and promoting environmentally friendly products. The company also began working on a Sustainable Products Index, aimed at increasing information on the attributes of products available to its consumers.[79]

The company's environmental sustainability attracted much positive press, yet many continued to criticise Walmart, alleging continued anti-union activities, relatively low wages, and lax supplier monitoring. The company began to receive recognition for its accomplishments, but many critics and detractors remained.[80]

Cases for comparison

Compare and contrast this case with Case 3 'Influence through voice and exit', Case 5 'Alleged versus confirmed illegal activity' and Case 9 'Incomplete societal and environmental data'.

Case 7

Public versus private partnerships for engagement

- **When is it appropriate to work with companies behind the scenes?**
- **When is it appropriate to take the debate public?**
- **When is it appropriate to engage in collaborative initiatives and actions with activist organisations?**

The case

You are a mainstream money manager with a major bank. You have a strong RI business with many clients. You are in private discussions with a large toy manufacturer with operations in China about its labour standards for its subcontractors in that country. The discussions with the company have been friendly. It has discussed frankly with you its problems and taken you to visit its factories. But it hasn't made much progress in improving its practices. You are approached by a high-profile labour activist non-governmental organisation (NGO) to join with it in a public campaign criticising the company. This NGO has often sharply criticised corporations' conduct in the past. You are tempted to join because you believe the company should be moving faster. But you would have to stop the private, confidential talks with the company to do so. You think that the best way to influence the company's management is through continuing quiet, private, behind-the-scenes discussions. But you wonder whether some harsh public criticism might not be more effective than your slow approach. You are tempted to join with the NGO, but you are not sure whether it will be counterproductive.

As you are relatively new to the firm and are not sure what to do, you have asked for a meeting with Chloé, head of the responsible investments team, to discuss the issue. Before meeting her, you want to set down on paper the approach you would recommend.

Which option would you favour . . . ?

Dilemma for the responsible investor

This case highlights dilemmas that arise when responsible investors engage both with corporations and with NGOs such as environmental or citizens groups on issues of societal and environmental concern. Responsible investors can play a valuable role in promoting societal dialogue with corporations and, under the proper circumstances, bringing together corporations and NGOs.[81] Determining what the right circumstances are and how best to use NGOs in communications with corporations is not always easy.

The case in perspective

This case focuses on how building trust between corporations and their critics can promote positive societal and environmental change, and on how that trust can be difficult to maintain. Mainstream activist investors occasionally band together to confront companies on issues relating to stock price. However, contending with societal and environmental concerns introduces challenges that require partnerships and dialogue built not so much on confrontation as on trust.

Building trust around societal and environmental issues can be difficult because corporate managers may need to be persuaded that such issues are a legitimate part of their management responsibilities. In addition, corporate executives may initially distrust non-profit critics, perceiving them as fundamentally hostile. Similarly, those coming from the non-profit world often distrust

→

Increasingly, however, NGOs and corporations are entering into collaborative dialogues or partnerships on controversial and difficult matters of corporate social responsibility (CSR), rather than establishing uncritical philanthropic relations or unbending antagonistic stances with each other.

The two parties have learned to talk more directly with each other, as both corporations and NGOs have grown in number, power and influence. According to the 2002 UNDP Human Development Report, in 1914 there were 1,083 international NGOs. By 2000 there were more than 37,000 – nearly one fifth of them formed in the 1990s.[82]

A form of stakeholder engagement, these partnerships have become an increasingly familiar feature of the CSR landscape and have been described by one academic as 'social problem-solving mechanisms among organizations'[83] that offer solutions that benefit both partners, as well as society at large. When successful, these dialogues and partnerships often leverage joint resources and capitalise on the respective competences and strengths of the corporate and NGO worlds. But partnerships are not always easy. Moreover, they can be resource intensive, often requiring the building of trust over months or years of negotiations.

Responsible investors can play a unique role in fostering corporate/NGO dialogues by bringing together these two groups that otherwise might not be willing to sit at the same table. They can be in a position to win the confidence of corporations and at the same time have the trust of major players in the NGO world.

These dialogues and partnerships pose risks, as well as hold the promise of rewards, for both parties. To be successful, they must capitalise on the so-called 'four Ps' of partnership: the purpose of partnership, the pact between the partners, the power relationship within the partners and the process of partnerships evolution.[84]

corporations and their for-profit motives. Responsible investors are often uniquely positioned to build trust with both parties and bring them to the table for productive conversations. However, maintaining the trust of both parties in circumstances where dialogues are difficult is often a challenge

As difficult as these dialogues may be, through trust building they can increase the understanding of what constitutes interests that are shared among corporations, investors and society. The deeper the understanding of these shared interests, the greater the chances of positive, cooperative change.

In particular, gaining and maintaining the trust of potential corporate partners and NGOs alike can take a long time and require considerable ongoing effort. Trust is often best built through behind-the-scenes dialogue. Going public with issues of concern too soon can derail the kind of dialogue necessary for building that trust with corporations. Conversely, refusing to go public in the face of negotiations that are going nowhere can undercut the trust with NGOs. If responsible investors are unwilling to criticise companies publicly, they run the risk that NGOs, their clients or the public more generally may come to distrust their sincerity.

Responsible investors therefore sometimes find themselves riding a fine line – they aren't exactly an NGO, but neither are they an unquestioning ally of the corporate world. They run the risk of losing the trust of one party or the other, or even both – and trust, which is difficult to win, is even more difficult to retrieve once lost.

Approaches available to the responsible investor

One of the key themes raised in this case is the difficulty of managing the public and private aspects of partnerships. It is often unclear when partnerships with companies and with NGOs will be more effective if they are conducted away from the public's eye or if they appeal directly and openly to the public. Moreover, partnerships with either corporations or NGOs run the risk of damaging the reputation of responsible investors if they appear to the public to be ineffective or mismanaged. The challenge is to find the right forum for dialogue with the right partners.

Approaches to these challenges generally vary along three dimensions:

- How public or private the dialogue is

- How active a role the responsible investor plays in the partnership

- What partners the responsible investor chooses to ally itself most closely with

Responsible investors have a wide range of options on how public an engagement might be. At one end of the spectrum, they can enter into a confidentiality agreement with a corporation and ensure that no material information about ongoing discussions be disclosed to others until both sides agree the time is right. At the other end, responsible investors can ally themselves directly with public critics of a corporation and insist that all dialogue be conducted in full public view. Where along this spectrum the tactic most likely to produce results lies will depend on the parties involved, the issue itself and the stage at which the negotiations find themselves.

Responsible investors also have a range of choices on the number of players with whom they choose to engage on both sides. Responsible investors can work independently, avoiding links to other responsible investors or NGOs, or form a loose or close alliance with several partners from the investment world or from among non-profit organisations. They can engage with a single company or with multiple companies within an industry confronting an identical challenge. Their role can be that of relatively passive facilitator, bringing the different parties together and letting a conversation evolve, or it can be that of a more active facilitator, directing dialogue towards specific ends.

Engaging with coalitions of corporations can be beneficial because of the prospects of industry-wide action. Engaging with coalitions of NGOs can bring similar benefits because they bring more weight to bear on the negotiations. Nevertheless, negotiations with multiple parties are likely to be more complicated and time-consuming. In addition, the need for privacy is likely to increase because the need to achieve consensus will require extensive behind-the-scenes negotiations.

Variable factors

The response to this case may differ depending on a number of factors, such as the type of financial institution the responsible investor is working with, the type of NGO that is approaching the responsible investor and the potential for the dialogue or partnership to create change.

- What type of financial institution is the investor working for?
    - Is it a mainstream institution that would be hesitant to associate with NGOs or is it a financial institution primarily focused on RI and consequently more comfortable with the NGO world?
    - Are the institution's RI products and services a large or small part of its overall business?
    - Does the institution have formal policies with regards to such collaborations?
    - What would the institution's clients think about the partnership?
- What is the nature of the NGO, its concerns and history?
    - Is it an NGO with a history of collaboration or confrontation?
    - Does the NGO have past experience with partnerships or collaborations?
    - Is the NGO well informed on the issue involved and does it have a clear goal with which it will be satisfied?
- Would the partnership help to create change at the firm?
    - Does the company have a history of working successfully in partnerships either on its own or in collaboration with other members of its industry?

- What is the company's attitude towards NGOs in general and specifically towards the NGO that might become involved in the dialogue?
- Are the general circumstances favourable for a partnership approach within the culture of the country, industry or moment in history?

Recommendations

As you approach dilemmas of this sort, you may want to:

- Conduct a risk assessment of engaging or partnering with an NGO
- Review possible alternatives before entering into a partnership in order to ensure that that is the most effective approach
- Be sure not to endanger the legitimacy of the organisation and its activities vis-à-vis companies, NGOs and its clients
- Make sure to know the NGO before partnering with it
- Set clear objectives and plans of action before entering into a partnership

Engagement interactions

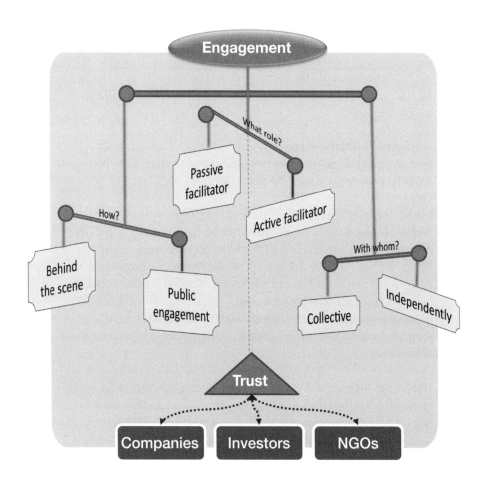

Responses from practitioners

Joining with an NGO may be the only solution . . . but not any type of NGO

NOTES TO MYSELF IN PREPARATION FOR THE MEETING WITH CHLOÉ

I believe that often companies will not change unless they feel strong pressure from outside. So it is useful to have an NGO to create pressure to change.

I can see arguments for criticising the company publicly but independently from the NGO or for joining the NGO's campaign. If we joined with the NGO, I would want to be sure they are credible and influential enough to be taken seriously by the corporation. I would also want to be sure that we are on the same wavelength as the NGO in our criticisms of the company. The decision really would depend on many factors relating to the particular situation.

The most important criterion is whether the NGO can present convincing evidence on the issue. If they are attacking the company, they need to back up their assertions with solid evidence. Our decision on whether to join them depends in large part on whether the NGO is professional enough.

Public engagement and joining an NGO may be important in this case because the company will most probably only respond to real pressures from outside. So I would definitely use public pressure.

Criticise, yes, but independently from the NGO

<u>NOTES TO MYSELF IN PREPARATION FOR THE MEETING</u>
<u>WITH CHLOÉ</u>

*I think that in this situation it would be best to criticise
the company's policies publicly but independently from
the NGO. The reason for this is the need for circumspection
and independence from other influencing bodies such
as NGOs. However, it would be quite proper to include
arguments and facts cited by the NGO in our independent
and public critique.*

> 66 The work that NGOs are doing informs the work
> that we're doing. There's potential for conflict, but
> there's also plenty of room for working together, aside
> from openly joining a campaign. 99

> 66 There is a fine line between working closely with
> the NGOs and keeping our dialogue with the company
> positive. It's definitely challenging . . . but we continue
> to engage with both sides, although more privately with
> the company. 99

Favour indirect collaboration to stimulate improvements

NOTES TO MYSELF IN PREPARATION FOR THE MEETING WITH CHLOÉ

Since I know that we never campaign against companies, either independently or with an NGO, we should probably:

a) Continue with our behind-the-scenes dialogue. This is not truly a private dialogue because I know it is our company's policy not to seek confidential information. We always keep open the option to publish the information we collect in our profiles, which are publicly available. If it is a question of confidential information, we ask the company not to disclose it to us. We can engage privately with the company but the information we seek should not be confidential.

b) Tell the NGO that we will not collaborate with them. It is their job to be activist, not ours. We want to stimulate companies to improve their sustainability performance, but not by publicly criticising them.

We could, however, join with the NGO in publicly raising one or two of the major issues that are of concern, without specifically targeting the company for criticism. For example, there might be a situation in which we want to work with a particular company in the pharmaceutical industry to increase its transparency on animal testing, or to increase its use of alternatives to animal testing. For the industry as a whole, we could raise publicly the issues of transparency and alternatives, but without naming the particular company with which we are engaging. That is different from campaigning against the company specifically. It is flagging the problem for public discussion, and could involve the NGO.

page 1

We could also take steps that involve the NGO but fall short of actual collaboration. For example, we could facilitate a meeting with the company and bring the NGO to the table, but only as a resource, not a partner. We could tell the NGO that we would like to consult with it during our dialogue with the company. We could use it to help establish certain criteria we are asking the company to meet, or as a source for the information we use in our dialogues with the firm. But again, these approaches would not really be a full-fledged collaboration.

> 66 We are often in a kind of balancing act between the activist world and mainstream financial community who don't tend to raise these questions – and there's a role to be played by being in a slightly different place but understanding the activist point of view and understanding how they help you, and how together you create change. 99

In the news

In 1997, the environmental advocacy group Rainforest Action Network (RAN) began a campaign against US company Home Depot to force the company to stop selling lumber harvested from old-growth forests.[85] At that time, the company was selling a number of lumber products harvested from old-growth forests, particularly from endangered areas in British Columbia.

RAN used a number of tactics to force the company to commit to phasing out sales of wood from old-growth forests, including high-profile demonstrations outside the company's headquarters, fighting corporate expansion plans at city council meetings, coordinating a national ad campaign, obtaining endorsements from high-profile celebrities, and organising demonstrations outside hundreds of Home Depot stores.

RI investors who owned Home Depot stock were faced with three choices: sell their stock in protest and support RAN's efforts, hold the stock and communicate with the company, or do nothing. Many RI investors chose to hold Home Depot stock and engage with the company to encourage it to adopt the new policy, but they remained under pressure to take more dramatic action from clients who were responding to RAN's negative public relations campaign.

The combined external pressure from RAN and internal pressure from shareholders was ultimately successful. As a result, Home Depot announced in 1999 that it would phase out and eventually stop selling lumber from old-growth forests. A number of other home improvement retailers then followed suit, helping to alter general practice within the do-it-yourself retail industry.

Cases for comparison

Compare and contrast this case with Case 2 'Ethics and facts' and Case 3 'Influence through voice and exit'.

Case 8

Relativity of responsible investment standards

- Are responsible investment standards absolute or are they relative to particular cultures?

- Can a responsible investment product be marketed globally or only regionally?

- Is there such a thing as a universal responsible investment standard?

The case

You are in charge of product development for a major European bank with worldwide operations. Your CEO, Mr Stephan, has just made a major commitment to develop RI products. You have been asked to design RI products that can be rolled out by your banking divisions and their marketing departments around the world in the next six months. You have taken as a model for this roll-out an RI product that your bank has already marketed very successfully in Europe. You just held a meeting of bank representatives from around the world to see how they think this European RI fund would sell in their local markets. You expected an enthusiastic response, but were surprised by their answers. The representative from the Middle East said that she absolutely couldn't sell the fund there unless it eliminated all financial services companies. Because the Koran forbids usury, Islamic RI funds cannot include banks. The representative from Asia said that the fund's requirement that women be represented on corporate boards of directors and in top management was totally unreasonable. Practically no Japanese companies had women represented in either place and such a requirement would eliminate all companies from investment consideration. The representative from South Africa pointed out that job creation and black empowerment were two issues that mattered greatly there. Emphasising the environment over these two issues was political and financial suicide.

What would you suggest to your CEO, Mr Stephan . . . ?

Dilemma for the responsible investor

This case highlights the dilemmas that arise when responsible investors attempt to impose the same set of RI standards on portfolios marketed in different countries or regions. The dilemmas arise because cultural and regulatory differences can cause certain issues to be high on the RI agenda in one region but not in another. This raises the question of whether RI standards should be viewed as culturally relative or universally applicable or, from a different perspective, whether responsible investors can apply different standards to different products without undercutting their legitimacy.

Traditional money managers can differ in their methodologies when it comes to their philosophies of investment. They may be growth managers, purchasing only companies that have a strong record of earnings growth even if their stocks appear expensive. Or they may be value managers, seeking out the undervalued stocks of companies that may have poor earnings today but represent under-appreciated long-term value. One could therefore ask if there is one approach to conventional investment that is universally applicable, or if investment styles work only at specific times or in particular circumstance. Or, from a different perspective, can a single money manager offer both these styles, which appear to a certain degree contradictory, and maintain his or her credibility? Large institutional investors tend to diversify across investment styles, on the grounds that each will function well under different conditions. Could the same argument be made for diversification of RI standards?

In part this dilemma arises because one of the purposes of RI is to communicate to companies' management clearly on matters of corporate social responsibility (CSR). It can be confusing to the managers of corporations if an RI firm takes a

The case in perspective

This case focuses on the question of whether societal and environmental standards for corporate conduct should be conceived of as global and therefore universally applicable, or as local and therefore variable depending on circumstance. It highlights the contribution that the responsible investment community has made to the evolution of the literally hundreds of industry-specific standards, codes of conduct and principles on a wide range of societal and environmental issues that now exist.

As these societal and environmental standards have proliferated, it is not surprising that they differ. People will have different views on the societal and environmental standards to which corporations should be held. It is also not surprising that questions arise as to competing standards' rigour, effectiveness and appropriateness. Understanding how different standards can be reasonably and rigorously applied in different regions inevitably raises challenges with which responsible investors must contend. It is in part through contending with these challenges that the standards for sustainable and ethical conduct by corporations can be raised around the world.

position in one region of the world that contradicts one that it takes elsewhere. For example, a fund that shuns nuclear power in one region because the nuclear waste disposal issue isn't yet solved, while in another it includes nuclear power utilities because they produce no greenhouse gases; or a fund that takes the position that having women in management is an important factor in one country but not in another.

At the same time, a regionally specific RI concern that may appear irrelevant to others can suddenly take on relevance as circumstances change. For example, after the worldwide financial crisis of 2007–09, the ban on investments in financial institutions imposed by Islamic funds suddenly appeared better reasoned and more relevant than others may have previously thought.

The situation is further complicated by the fact that the concept of CSR is interpreted differently around the world, with conflicting views as to its nature and no generally accepted standard.[86] The concept of CSR is increasingly winning worldwide recognition. An ever-increasing number of standards and principles for corporate social and environmental conduct are being developed around the world. Yet the implementation of CSR varies from country to country and from culture to culture. It is a practice that can be described as simultaneously global and local.

In addition, one need not look across regions to find differences in the application of RI standards. Within a single country or region, the same provider of RI products might offer a product strictly screened on all societal and environmental issues for 'deep green' retail clients and a second product where only nuclear weapons manufacturers are excluded for the institutional investor market. Or, one large pharmaceutical company might be screened out by one RI fund and included in another. Critics of RI occasionally perceive this type of relativity as a weakness of RI, while proponents present this diversity of viewpoints as a strength.

This same dilemma can play itself out in the marketing of RI funds. Marketing universals can be defined as consumer behaviour within a segment or particular product category that does not vary across cultures.[87] It is not clear, however, that such universals can be applied successfully to RI products. That is, if RI products are marketed as having universal appeal, they will almost certainly risk raising complicated issues of cultural norms. Conversely, if they are targeted to specific regions, they may forgo the possibility of finding universal appeal.

Asserting universal standards has certain clear advantages. It can improve international co-ordination in the RI field, facilitate interaction among RI groups, enhance the compatibility of RI efforts and encourage clear communications with corporate management. But standardisation also runs the danger of cutting off fruitful debate that can lead to the incorporation of diverse and useful points of view, to the consideration of emerging issues, to the recognition of innovation in the field and to the promotion of a variety of steps forward on complicated issues.

Approaches available to the responsible investor

One of the key themes raised in this case is how to approach the tensions between the local and global that inevitably arise in RI. Three general approaches suggest themselves:

- Fully adapt the products' RI criteria to local and regional considerations

- Construct an RI product with a single set of standards and methodologies that are applied consistently to all regions

- Provide a mixture of RI products with different criteria set to different tolerance levels that can be tailored to a variety of locally accepted standards

The decision on which approach to take will vary depending on the type of institution involved, its general marketing strategies, the resources available to it and the manner in which it conceives of RI.

A financial institution with global reach and considerable resources might be naturally inclined to adopt an RI approach tailored to specific regions. Its global operations would give it sensitivity to the variety of issues in specific regions, while the large scale of its operations would allow it to launch and market successfully a variety of different products. By contrast, a smaller firm with operations in a single country would probably be less inclined to create a variety of funds serving a variety of RI markets.

If RI is only one of many products offered by a financial services firm, or if the firm's business strategy has historically been to offer a multiplicity of customised products, offering a suite of RI products with a variety of RI standards would be a relatively natural option. On the other hand, if the company's products were focused solely on RI and it took prominent, public stands on specific societal and environmental issues, it would be more difficult for it to launch a variety of products with a variety of differing standards.

In confronting this dilemma, financial services firms needn't necessarily choose one of these options to the total exclusion of the others. It could, for example, establish a set of minimum standards that applied to all its products, but leave room for local adaptation in regionally marketed funds on other criteria. Or, it could establish general themes that would be universal across all funds, but their application would be adapted to regional circumstances. For example, diversity in the workplace might be established as a general criterion for all the firm's funds, but what specific criteria were used to evaluate firms would differ by region. In one country, the focus might be primarily on how successful companies had been in incorporating women into senior management. In another, the distinguishing feature might be the percentage of employees with disabilities. In a third, the key factor might be support for diversity in sexual orientation. In some regions, progress in achieving work–life balance might be well established through national regulation and therefore not a crucial issue for corporate management, while in others it might remain a substantial challenge on which companies could be meaningfully evaluated.

Whichever approach an RI firm takes, transparency and clarity on methodologies will be crucial. The firm should be clear whether the justification for the standards it adopts is global or local. It should be clear on the norms that underlie these standards. More generally it should provide a vision of the society towards which, as responsible investors, the firm sees itself working. The standards, whether local or global, are in effect no more than the means of working towards that overall vision.

Variable factors

Responses to this case may differ depending on a number of factors, such as the *raison d'être* of the RI fund, the regions targeted by the fund, the impact of criteria on the composition of the fund and the type of RI standard-setting used by the financial institution in question.

- What is the purpose of the RI fund?
  - What is the RI fund trying to achieve?
  - Is it focusing on specific societal or environmental goals?
  - What is the core market for which the fund has been created?
  - What are the resources available to the institution?

- What are the regions targeted by the institution?
  - In which regions are the principle clients of the RI fund located?
  - Are RI issues viewed similarly throughout these regions?
  - Do these regions have specific values and norms that must be incorporated for the fund to be successful?

- What is the impact of the criteria on the composition of the portfolio?
  - Are major industries or companies excluded because of locally relevant RI criteria?
  - Do the regional or global RI criteria permit the inclusion of the major companies or industries that make up the economy of the region?

- Is the fund based on an exclusionary, positive or best-in-class approach?
  - How commonly accepted are the exclusionary criteria that the firm has already imposed on its existing RI funds?
  - How strictly does it impose these criteria?
  - Is the fund's fundamental philosophy a values- and principles-based one or is it oriented towards value and risk control?
  - Does the firm have a set of minimum standards that it is committed to applying to all its RI products?

Recommendations

As you approach dilemmas of this sort, you may want to:

- Make explicit what the firm wants to achieve with its RI products
- Explain the purpose and approaches of RI to sales managers
- Determine if the institution has an explicit or implicit set of minimum standards that it applies to its RI products
- Make sure that the standards developed do not lead to the exclusion of all the major companies in a specific country or area
- Know and understand the expectations, norms and values of the firm's clients
- Make a distinction between criteria that are more widely and less widely accepted at an international level

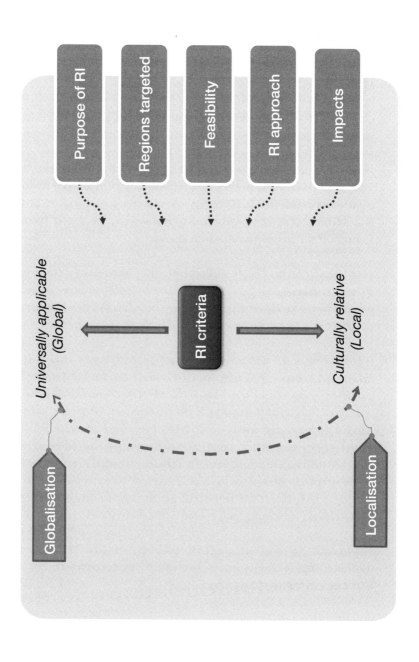

Responses from practitioners

One RI fund everywhere in the world

NOTES IN PREPARATION FOR THE MEETING WITH MR STEPHAN

Our objective is to sell the same RI fund everywhere in the world, or at least to the maximum extent possible. For that we need to educate our people throughout the bank so that they can explain what this product is all about and why we have chosen certain criteria. We must try to have objective criteria. Just as financial analysts use essentially the same criteria throughout the world, we want to do the same with non-financial criteria.

The local context will nevertheless remain very important as we analyse companies for our portfolios and consider what they are doing in specific countries. It is the local context for the company analysis that is important.

We may end up with a fund that is sold everywhere: US, Europe, Dubai. This fund will have the same standards in all these countries and we will not try to make it country specific. Our challenge will be in convincing our sales managers in different countries how best to market it locally. For example, in Austria we will need to be clear why we allow in companies with some small involvement with nuclear power, while in France we won't need to explain this particular point. Such involvement would be considered normal there but not in Austria.

Explanation and education will be crucial. The way we explain the criteria may change from country to country, but the criteria stay the same.

Regional adaptation of RI products

<u>NOTES IN PREPARATION FOR THE MEETING WITH</u>
<u>MR STEPHAN</u>

A wide range of RI approaches exits throughout the world.
I would not expect one RI fund to be optimal for all
regions. We need to be a player in various regions in order
to know how to adapt our products to these differences.

For example, although the environment is an issue
applicable everywhere, Japanese investors are particularly
demanding on environmental criteria. Environment
is one of the most important issues there. Specific
issues should be considered on a local basis. Issues
such as Islamic finance and employee or community
empowerment are very country specific.

→ Create different funds for the different regions. A
   local approach to RI is necessary.

But I would say it is not just country specific but, and
above all, specific to the type of investor. Church and
pension funds can take very different approaches, even
within the same region. So it does not really depend on
where they are from but rather on who and what type of
clients they are. We should be thinking of our funds more
as specific to the type of investor we want to attract, rather
than to the region in which we want to market them.

→ A global approach to RI would be very difficult.

Key deciding factors: the criteria and the threshold

<u>NOTES IN PREPARATION FOR THE MEETING WITH</u>
<u>MR STEPHAN</u>

The recommendations I would make will in part depend
on answers to a few questions.

● <u>What is the reason that we have this fund?</u>
a) Is it a purely marketing-driven exercise - we are
going global because there is a market out there and it
is simpler to have one product? Or,
b) Is our objective to achieve particular societal or
environmental goals and therefore we may want to have
particular emphasis on certain criteria over others?

● <u>What is the level of demand for the products we have in</u>
<u>mind?</u>
To answer this question we need to do some market
research to determine if we want to shape the fund or
funds in a particular way. The question here is how
relevant is the relativity of RI standards. If there is not
'one' standard for RI, then there may be many different
ways in which we can approach this market.

> We need to be very transparent about methodology and
> approach, and the reasons why a particular approach
> has been chosen.

As to the question of global versus local RI, I would say
this is a challenge for many funds. Most fund managers
think of launching global funds but thus far you'll find
very few, if any, global RI funds. You might find funds
sold across Europe, but you won't generally find funds
that are sold around the world, because these differences
in criteria exist. In particular, it's highly unlikely that
Sharia or Islamic finance would be compatible with
other forms of RI. I think Sharia is a form of RI. But the

page 1

difference is, if you are setting up a Sharia-compliant fund, it has a different set of criteria and could not be a default RI fund outside the Islamic world.

So if you are trying to sell a fund in different countries there are two approaches you can take:
a) You can have one fund that has a relatively common set of criteria. It isn't difficult to establish a series of environmental questions, or even societal questions, that are of common concern around the world. But there may be criteria that are regionally specific, which you then would leave out of that core fund. Or,
b) You could set up a family of funds where there would be one core fund that would contain a cross-section of criteria that are common everywhere and then have subsidiary funds based on local markets.

> It depends how you define the criteria and your thresholds.

It's true that few Japanese companies have women on their boards of directors. That's a challenge for Japanese companies. If you make it a criterion that excludes all companies that don't have women on the board, of course you will not be able to include many of the Japanese companies that play an important role in that country's economy. But if you have a criterion that is a bit weaker than that, you might be able to incorporate a version of this criterion.

> Distinction between regions

In the end you don't want too many country-specific funds. But you could have an Asian, a European one and a North American one, I don't think you need more than that.

page 2

> Distinction between clients

But you may also have to make a distinction between retail and institutional funds. If it's a retail fund, excluding nuclear power may not be a problem in many countries. It's not going to stop our firm from selling a fund. But if it's an institutional fund, a nuclear energy exclusion might become a problem.

You can have a global fund, but if you do that, you just have to make sure that you don't have too many screens that are controversial in particular market places. You can have a core fund, but it might have a more limited set of criteria. That might be sufficient for a globally marketed fund. It would meet enough basic RI expectations in most parts of the world, especially on environmental issues, but also on certain societal issues, that it could be marketed globally. If you want to have it really 'dark green' with exceptionally strict environmental criteria, you could still do that separately, but you would want to be sure to have a core fund that is 'lighter green' as well.

66 RI means different things in different parts of the world. It is not that there are only regional differences but there are different types of clients in different regions. Clearly it is very difficult to have one RI product that fulfils these different needs. 99

66 Part of the issue is about understanding the purpose of RI. In that regard the funds should be culturally sensitive but also seek to address challenging issues and to push the envelope on certain issues. 99

In the news

In practice, investors worldwide have a range of RI different priorities, informed in part by varying cultures and standards. These priorities affect the way they interact with their investments and their definitions of what societal responsibility means.

For example, Islamic finance – or finance that is carried out in compliance with Islamic Sharia law – is one of the world's fastest growing segments of the financial system, with around $822 billion in Sharia-compliant assets at year end 2008.[88] For Muslims, the Koran can be used as the basis for establishing criteria for how they invest. The Koran can be interpreted to prohibit investment in companies involved in usury (*riba*, the collection and payment of interest) as well as those that contravene other Islamic principles (for example, companies that deal in pork, alcohol and pornography). Similarly, it can be interpreted to encourage investors to emphasise profit- and risk-sharing and the safekeeping of assets. These religious principles have come to shape and form a philosophy of RI investment that many Muslims are increasingly adopting.[89]

Cultural heritage and historical or economic realities can have an impact as well. In South Africa, one of the legacies of the struggle against apartheid and the exclusion of that country's black population from the benefits of the South African economy has been to elevate to the top of the RI agenda there issues of poverty alleviation and the inclusion of blacks in the ownership and management of corporations. A number of RI investment funds there have been founded with the explicit purpose of reducing poverty. Government efforts to encourage black economic empowerment have resulted in the passing of the Broad-Based Black Economic Empowerment Act of 2003 (BBBEE), which in turn has served as a new measuring stick for investors interested in evaluating companies' records on the just and sustainable development of the country.[90] Various provisions of the BBBEE make it possible to score individual companies' progress in including blacks in corporate affairs. All South African businesses are strongly encouraged, but not required, to fill out a BBBEE scorecard that can be used to assess aspects of their social performance. BBBEE performance is also a component of the Johannesburg Stock Exchange (JSE) Socially Responsible Investment Index.[91]

Cases for comparison

Compare and contrast this case with Case 1 'Types of responsible investor'.

Case 9

Incomplete societal and environmental data

- **What should RI investors do when faced with incomplete societal and environmental data?**
- **Can an evaluation be made with information that may appear incomplete?**
- **Should smaller companies with less societal and environmental information available be assessed differently from large companies?**

The case

You are the head of the research department of a major European bank. Over the past month one of your sustainability research analysts, Sarah, has been reviewing the corporate social responsibility (CSR) records of 30 European software, semiconductor and electronic equipment companies.

Ten of these companies are large, well-known firms. They have complete CSR reports. Their CSR activities and major controversies have been well covered in the press.

Ten of these companies are medium-sized firms that have published a moderate amount of CSR information in their annual reports and you are reasonably confident that any major CSR controversies have been covered in the press.

Ten of these companies, mostly smaller firms, have not published any CSR information and your press searches uncover no information – positive or negative – on them. Sarah has contacted these companies directly, but they either have CSR programmes but don't report publicly on them or they have not responded at all. You are not certain that controversies about the companies would have been reported in the press. Moreover, you know from an informal source, that one of these companies has taken some very innovative CSR initiatives recently.

Sarah does not know what recommendation to make about these companies. She comes to you for advice with regard to the last group of ten companies, the ones that do not provide any information.

What would you advise Sarah to do . . . ?

Dilemma for the responsible investor

This case highlights dilemmas that arise because, for a variety of reasons, smaller companies tend to report less frequently and thoroughly on their societal and environmental records than their larger industry peers. In addition, they are generally less well covered in the press. The first challenge faced here by RI analysts is whether it is possible to obtain additional data. If it is not possible, then the second challenge is whether the smaller companies should be penalised for the absence of data – positive or negative – on their societal and environmental records. The dilemma is how to proceed in a way that is fair to both large and small companies, while at the same time acknowledging that more data is needed from the smaller firms and encouraging them to disclose more.[92]

Mainstream money managers often face similar situations with regard to financial data. Larger companies in various industries tend to be covered better by Wall Street analysts and the business press, and management of these larger firms often communicates more frequently with the financial community. For financial analysts to communicate with and visit these smaller firms is an expense and commitment of time that makes it difficult for them to follow smaller companies in substantial numbers.

Over the past 20 years, an increasing number of the larger firms globally have begun issuing CSR reports. According to KPMG's *International Survey of Corporate Responsibility Reporting*,[93] nearly 80% of the largest 250 companies worldwide issued CSR reports in 2008. For a variety of reasons, one might expect larger firms to begin reporting before their smaller counterparts. They tend to be higher-profile

The case in perspective

This case focuses on responsible investors' continual search for more, and more detailed, CSR-related data. It is not an exaggeration to say that one of responsible investments' most important accomplishments has been to help promote the remarkable growth in CSR reporting around the world since the early 1990s. Although, as of the first decade of the 21st century, the mainstream financial community made relatively little use of this data, for responsible investors it was, in some senses, never enough.

Responsible investors often find the CSR data available inadequate for their needs; in part because CSR reporting is voluntary and consequently many companies simply do not report; in part because companies don't report in standardised formats, making comparisons difficult; in part because companies report well on some issues, but poorly on others; in part because data reported can be difficult to analyse without additional context; and for a host of other reasons.

→

companies with larger societal and environmental footprints. They are subject to greater scrutiny from their stakeholders, governmental bodies and the press. Their size gives them greater resources to implement, track and report on their CSR performance.

In addition, practitioners and academics have conducted most of their work on CSR with large corporations in mind. The current set of tools for measuring and reporting on CSR may simply not be well designed for smaller firms. 'A small business is not a little big business'[94] and in many senses the characteristics of large and small firms are not comparable.

Nevertheless, the RI community's demands for more and better CSR reporting have resulted in major progress in corporate management of many issues of crucial societal and environmental concern. How to obtain additional CSR data and how to conduct research day to day in the face of the desire for more such data are challenges that are likely to be with responsible investors for some time to come.

Finally, the public in general focuses on large companies when it comes to CSR issues, so smaller firms are under relatively less external pressure to report. Internally these firms may not have addressed CSR issues formally, or, if they have, they may not have the time or resources to create a stand-alone CSR department to measure effectiveness and report on implementation.

Because smaller companies are less well covered in the press, controversies at these firms may not be known to the outside world. In these circumstances, a company may be reluctant to raise controversial issues publicly itself. Moreover, environmental activists or community groups tend to target the larger firms in an industry because their names and brands are better recognised. Targeting smaller companies would receive less coverage and has less impact on the industry as a whole.

For these reasons, an RI analyst may not know if behind a lack of societal and environmental data for a small company lie unrecognised controversies or unreported positive initiatives. An absence of information could also mean a neutral record, with no particular positives or negatives relevant to the firm at all.

Because of the ambiguous significance of a lack of societal and environmental data, it can be difficult for RI analysts to decide what constitutes a fair judgement about smaller firms.

Approaches available to the responsible investor

One of the key themes raised in this case is that of decision-making in the absence of complete information. This dilemma reflects some of the larger challenges that CSR analysts face with regards to the comparability, consistency, reliability and comprehensiveness of CSR information in general. When facing such situations, responsible investors can take a number of approaches:

- Seek further information from firms where information is incomplete

- Provide provisional ratings until more information is available

- Apply a transparency criterion, downgrading a company that is not fully transparent

- Give the company the benefit of the doubt, assuming that no information implies no controversies

The first option – seeking more information from the company itself and from its stakeholders – is a likely approach in such situations. Direct contact with smaller companies can yield stories about CSR initiatives that the company has not previously disclosed. Contacts with stakeholders – environmental groups, community activists, unions or local newspapers – can unearth controversies or confirm positives.

However, additional research does not always produce definitive results – companies don't always cooperate, and the appropriate stakeholders can be difficult to locate or can themselves only provide incomplete assessments. It also necessitates expenditure of time and resources that may not be available to the responsible investor.

Responsible investors can also penalise companies that do not disclose their CSR records on the grounds that size should not be used by companies as an excuse for not monitoring and reporting on CSR issues. This approach is straightforward, but runs the risk of excluding companies that may have positive or neutral records. To exclude a firm because it does not have the resources to do formal CSR tracking and reporting can be unfair and also eliminate companies that could be beneficial to clients' investment portfolios.

To give the benefit of the doubt to smaller companies for which no negative information shows up in the press or from NGOs assumes that if controversies were truly major, they would have been brought to the public's attention. A lack of disclosure of positives by a company is not by itself sufficient reason to exclude it. The approach involves what is in effect a provisional rating – a rating that is reasonable given the best information currently available.

To address this disparity between larger and smaller firms, the Global Reporting Initiative (GRI) has established a simplified CSR reporting model specifically designed for small companies.[95] The amount and type of information required from these small companies is less than that from their larger counterparts. The GRI's reporting regime acknowledges the fact that CSR processes in smaller companies can be less formal and their resources for monitoring and reporting scarcer.

Another layer of complexity is currently being added by the fact that some companies are experimenting with different disclosure techniques, including the use of electronic and social media and, in doing so, abandoning printed CSR reports. Still others are developing means of reporting to consumers as well as investors. This proliferation of reporting media and methods increases the chances that responsible investors will be working with differing amounts of information for firms, both large and small.

Variable factors

The response to this case may differ depending on a number of factors, such as the resources available to RI analysts for further CSR information-gathering, the weight given by the analysts to CSR reporting relative to actual performance, the nature of the core business of the non-disclosing companies and the nature of the portfolio for which the companies are being considered.

- What resources can the RI analysts devote to pursuing additional information?
  - How much time and how many resources can the organisation devote to gathering additional CSR data for smaller companies?
  - What are the relative costs and benefits of devoting resources to look for additional CSR information?
  - Are the methodology and criteria for assessing smaller companies realistic in light of the information available?

- How important do the analysts consider the fact of reporting itself?
  - What importance does the organisation place on CSR programmes and performance as opposed to reporting on these programmes?
  - How much information on CSR programmes and performance is required by the organisation's methodologies in order to make a judgement?

- What is the core business of the non-disclosing company?
  - Is the company in a high societal or environmental risk sector such as oil, forestry or mining?
  - Is the company in a sector that has a high sustainability value such as alternative energy or green chemistry?
  - Is the company in a country where the press or governmental regulators are likely to have publicly reported on controversies as they arise?

- What is the portfolio's objective and structure?
  - Is the portfolio sector specific and does it therefore require representation from smaller companies in that sector?
  - If it is a multi-sector fund, what proportion of each sector is likely to be made up of smaller companies?
  - What percentage might non-disclosing companies represent of the total portfolio?
  - Should one of the objectives of the fund be to influence smaller companies to disclose more CSR data?

Recommendations

As you approach dilemmas of this sort, you may want to:

- Be certain to have all information that is reasonably available given the time and resources available

- Distinguish between situations in which information on companies is incomplete and situations in which no substantive positive or negative information is to be had at all

- Encourage greater disclosure by smaller firms

- Ensure that decisions on companies for which the information in hand is minimal are consistent with the fund's stated methodologies and are made consistently

The information availability challenge

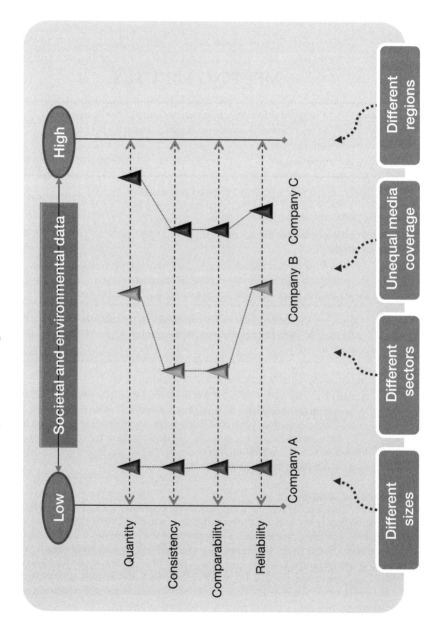

Responses from practitioners

Ensure a good selection process

MEETING MINUTES

Subject:	European software, semiconductor and electronic equipment companies: evaluation of companies which do not provide information
Attendees:	Bill, Sarah

We agreed on the following in our meeting this morning.

<u>For now:</u>
The nine companies that did not provide any information should be **kept out of the portfolio**.

> *Action:* Write to the companies saying we have not yet included them in the portfolio because of an absence of information on their CSR initiatives. Inclusion will be considered if they provide additional information.

> <u>Objective:</u> Create an incentive for companies to provide more information. Therefore: exclude them, but inform them and give them the chance to respond.

<u>The one company for which we have informal information:</u>

> *Action:* Look for sufficient proof that the information is trustworthy. We need to be certain that our clients have confidence that we are investing on the basis of reliable information. We would like written documentation. Based on the informal information we can go to the company and ask for confirmation and further details. We will inform the company that we are thinking of including them in our fund, but we need the additional information.

<u>Underlying principles</u>
This decision is in keeping with our underlying principles for investment. The question here is not so much the rating we assign to the company, positive or negative. The key is to enter into a dialogue with the companies to encourage them to be more forthcoming. Of course, they won't all end up with positive ratings and be considered for the fund. We should also keep in mind that our RI fund aims at identifying companies that are strong in financial terms, as well as in societal and environmental terms. The strongest in both areas are the ones we should be choosing.

- page 1 of 2 -

<u>Large versus small companies</u>
Our research focuses on the societal and environmental impact companies have. The question is: Is a given firm a high-impact or a low-impact company? Their impact is in part proportional to their size, but basically small and large are evaluated the same way.

There is one area where smaller companies can excel. A small company may be better able to innovate and think outside the box, coming up with programmes that are particularly positive, or addressing societal or environmental challenges in unusual ways. Here we wouldn't be concerned if they don't have particularly rigorous systems for measuring impact or formal policies in place. What we are looking for are examples of initiatives that larger companies, more concerned with benchmarking and standardised reporting, may not have thought of.

> 66 I think it is really important to make an effort to understand what is going on in smaller companies, and therefore we should focus on access to companies and actually meeting with them to fill in specific gaps. 99

> 66 As a responsible investor one of my roles is to try to increase the future availability of information. So, with that in mind, I would say that when there is no information, I will communicate to the company that this is a negative. By taking this approach, I can perhaps increase the amount of information that will be available in future years. 99

> 66 In an ideal world it would be great if everybody would produce a CSR report. But it would be naïve to expect smaller companies to be able to put as much time and resources into those types of things as the larger firms. 99

No disclosure is symptomatic of potential problems

MEETING MINUTES

Subject:	European software, semiconductor and electronic equipment companies: evaluation of companies which do not provide information
Attendees:	Lynn, Francis

> Disclosure is very important. Transparency is the cornerstone of CSR. Basically, if a company is not disclosing information, if a company is not transparent, there is probably a problem.

Argument

Some types of disclosure are easy to achieve, including environmental policies, values regarding CSR, and equal opportunity hiring and promotion policies. Disclosure in these areas is important because by publishing its values and policies with regard to CSR a company becomes subject to outside evaluation and is increasingly forced to engage with its stakeholders. Disclosure is a powerful means for encouraging self-regulation and can lead to continuous improvement. It pushes companies to go beyond the basic legal requirements.

➤ Therefore: **disclosure is crucial and essential**.

If no information on the 'easy issues' (basic environmental policy, equal opportunities, whistle blowing, etc.) is available, then the company should be excluded. This applies to small and large companies.

Of course, there are degrees of disclosure and many different types of information. For example, we would not require small companies to report on their CO_2 emissions because it is not easy to get this information. But basically, we require a minimal level of disclosure on their policies on carbon emissions.

To take into consideration: potential reputational risks for us

We would also ask if there is a high risk to the integrity of our fund in investing in this company if we are not sure about its performance or if we have incomplete data. If the risk is high, we would not invest. If the risk is low, even if there is incomplete data, we might consider the company, giving them the benefit of the doubt.

Conclusion

Transparency is crucial and essential. If no information is available on a company we cannot afford to invest in it because that would be against one of the basic concepts of our RI funds and because of the implicit reputational risks. But if the company discloses some basic information on policies and if the risk to the integrity of the fund is low, we may give them the benefit of the doubt.

In the news

Corporations can differ substantially in their internal implementation of CSR policies and practices, as well as in the extent to which they monitor and report on them. The following example shows the range of information availability within a single industry – technology. As of May 2010, two high-tech companies, IBM and Dell, reported extensively on a broad variety of CSR programmes. Two others, Avid Technology and Logitech, reported on little.

At that time, IBM's website contained a number of pages pertaining to the company's CSR performance and initiatives, from the company's overall CSR strategy to specific information on employee, governance, public policy, community, supply chain and environment programmes. The website also included the company's CSR reports from 2002 to 2009, with a GRI Index in the most recent report.[96] Dell's corporate responsibility site included a number of sections that covered the company's CSR policies and performance in areas called 'Dell Environment', 'Dell Difference', 'Diversity', and 'Corporate Accountability'. Each section had multiple pages of information. In addition, Dell separately provided its most recent CSR report, complete with a GRI Index.[97]

At the same time, CSR data on Logitech's website was confined to details on its corporate governance, which included Board of Directors, committee charters, corporate governance principles, the company's code of ethics and the company's societal and environmental responsibility policy. There was no information on the implementation of the company's CSR policies.[98] Avid Technology had even less CSR information. Its website featured only a statement on the company's corporate governance structure with a few examples of its commitment to corporate governance standards. Investors looking for societal and environmental information from these two companies' public documents would have found themselves substantially at a loss.[99]

Cases for comparison

Compare and contrast this case with Case 2 'Emotions, ethics and facts', Case 3 'Influence through voice and exit', Case 6 'When a company changes' and Case 11 'Emerging issues'.

Case 10
Exclusion of industries

- Is it appropriate for responsible investors to eliminate entire industries?
- What should be the motivation or justification for such exclusions?
- How strictly or loosely should these industries be defined?

The case

You are the head of a sustainability team at a major insurance company. Your philosophy to date has been to use a best-in-class approach to RI – that is, your company does not eliminate any industries as a whole from consideration for investment, but rather selects the best in each industry for investment.

The CEO of your company sends you a memo asking whether the company should adopt a policy of not investing in manufacturers of landmines and weapons of mass destruction including nuclear weapons. She points out that a number of major pension funds, RI indexes and RI managers (including some of your competitors) have adopted such a policy.

She also wonders if uranium-mining companies, nuclear power companies or other industries should be excluded as well.

She presents five alternatives. They are:

- Continue the current policy of not eliminating any industries on the basis of their basic business

- Adopt a policy of not investing in companies that manufacture landmines and weapons of mass destruction including nuclear weapons

- Adopt a policy of not investing in companies that manufacture landmines and weapons of mass destruction and also not investing in uranium-mining companies

- Adopt a policy of not investing in companies that manufacture landmines and weapons of mass destruction, uranium-mining companies and nuclear power companies

- Adopt a policy of not investing in companies that manufacture landmines and weapons of mass destruction, uranium-mining companies and nuclear power companies, and consider not investing in tobacco companies as well

Which of these five alternatives would you recommend and why . . . ?

Dilemma for the responsible investor

This case highlights the dilemmas that arise when responsible investors choose to exclude certain industries from their portfolios.[100] These dilemmas arise because the societal or environmental implications of certain products, such as landmines, nuclear weapons or tobacco, appear to be antithetical to the spirit of responsible investment. However, responsible investors must confront three questions if they choose to avoid whole industries: (1) On what specific basis do they justify excluding entire industries? (2) How do they define involvement in the industries they choose to exclude? (3) What costs – either financial or in the ability to influence – might they incur?

The case in perspective

This case examines the practice of some responsible investors of excluding whole industries on the basis of societal or environmental criteria. Responsible investors do so for a wide range of reasons – they may view these industries as unethical, unsustainable, as posing unacceptable financial or reputational risks, and so on. It is also possible to view this practice as an assertion that governmental action, rather than market mechanisms, is necessary to address the substantial challenges posed by these industries. There is, for example, little point in communicating with corporations on the need to dismantle the nuclear weapons arsenals of the world or to address the problem of the health effects of second-hand smoke. Only government action, not market mechanisms, can address issues such as these. By contrast, mainstream investors generally view such limits

➜

Many mainstream investment disciplines involve the exclusion of certain classes of stocks or stocks in certain industries. Money managers who take a value approach frequently exclude companies that have a ratio of price to earnings above a certain level and therefore appear expensive. Conversely, growth managers exclude from consideration companies that don't have a record of above-average growth in earnings. Emerging markets funds exclude companies domiciled in developed countries. Small cap managers exclude companies above a certain size. Managers of sector funds concentrate on single industries (e.g. healthcare, high tech) and exclude all others. Virtually any money manager who doesn't invest in an index fund covering the whole market by definition makes some exclusions.

To these mainstream investors, however, excluding industries for societal or environmental reasons appears to be qualitatively different from their practices and, at least in theory, costly. The financial argument against exclusion is based on the assertion of modern portfolio theory that any restriction in the ability to diversify necessarily limits the ability of managers to maximise risk-adjusted returns. The question of whether industry-based exclusions hurt RI portfolios in practice has been debated by academics and practitioners for the past 30 years, essentially without resolution.

The theoretical arguments date back to the 1980s. Rudd, for example, asserted in 1981 that environmental and societal criteria would increase investment risks because of a detrimental and permanently negative impact of these criteria on the level of portfolio diversification.[101] This theoretical argument has found wide acceptance within the financial community, although some argue that this debate about theory is still unresolved. Numerous academic studies of actual performance have also been published over the past 30 years. Some have found outperformance by RI funds, some under-performance, and many no statistically significant effects at all. The RI index with the longest track record, the MSCI KLD 400 Social Index, outperformed its analogue, the Standard & Poor's 500 Index, over the first 20 years of its existence from 1 May 1990 to 1 May 2010.[102]

In addition, critics of exclusionary approaches raise two basic non-financial caveats about exclusionary criteria. The first relates to the moral justification for exclusions – that is, they ask whether such

on their ability to diversify as potentially detrimental to their risk-adjusted returns and consequently oppose exclusions of this type.

Deciding when to avoid certain industries and for what reasons can be a challenge for responsible investors. Such exclusions can alienate prospective clients who disagree about the need for action on certain issues, or institutional investors who don't wish to limit their opportunities for diversification. In addition, the decision not to invest represents a loss of influence because communicating with companies as a shareholder is no longer possible. Furthermore, the problems posed by these industries are in many cases likely to be with us for years, if not decades, to come.

Debates about whether to exclude certain industries – and if so, which ones – will continue to serve a useful purpose within the responsible investment world. They allow investors not to profit from certain activities they view as unethical or unsustainable. They also focus the public's attention on the need for governments, rather than markets, to address certain industry-level challenges.

criteria don't simply represent personal and idiosyncratic viewpoints that may be held by a limited number of investors, but are not widely enough held to justify imposing them on all investors in a fund. These critics point, for example, to a fund that excludes alcohol firms and ask if substantial numbers of investors in today's society truly believe that alcohol consumption should be avoided. In addition, they make the practical argument that if the goal of responsible investors is to promote corporate change, they are working against their own goals by forgoing the very ownership that gives them the right and opportunity to engage with management in their role as stockowner.

Moreover, if responsible investors believe they are justified on both financial and non-financial grounds in adopting an exclusionary approach, they then must decide to which industries to apply this approach and how to define involvement in those industries they have singled out for exclusion. The latter – questions of definition of involvement in specific industries – can in practice be more difficult than it might appear at first glance. If a responsible investor chooses to avoid manufacturers of nuclear weapons, does that definition include only the manufacturers of a nuclear warhead or does it also include manufacturers of delivery systems such as the missiles or submarines through which these warheads are launched and whose sole function is the delivery of these weapons? Should the definition be extended even to include aircraft that might deliver nuclear weapons? Or to the military satellite systems essential for the guidance of the missiles that deliver the nuclear warheads? Similar questions of what should be defined as involvement arise for virtually every industry that might be declared ineligible for investment.

Approaches available to the responsible investor

One of the key themes raised in this case is that of justification for the application of societal and environmental standards to whole industries. Once responsible investors have decided to adopt an exclusionary approach, they must justify it on grounds that are:

- Financial
- Ethical
- Legal
- Practical

On the financial side, this justification must assess the implications of the restrictions imposed for the returns of the fund. Because there is no simple answer to the question of whether exclusionary approaches hurt (or help) financial performance, these policies and practices must be examined case by case. The two most important factors here are the number of industries excluded and the number of companies excluded within each industry. The greater the number of industries and companies excluded, the greater the chances that the standards may increase the volatility of the fund's returns and adversely affect its risk-adjusted returns. The fewer the exclusions, the less likely it is that there will be negative effects.

Another consideration is the investment style of the fund – that is, is it a growth or value fund, domestic or international, large capitalisation or small capitalisation? Certain exclusionary policies will cut more deeply within certain types of approaches. For example, a value fund or a fund investing solely in Canadian or Australian stocks might be heavily represented in cyclical natural resources companies. The imposition of exclusions on major areas of natural resource extraction could eliminate large portions of the investable universe.

On the moral or ethical side of the equation, a number of different justifications for adopting an exclusionary approach are possible. Justifications coming out of the religious community can be strictly moral – that is, certain products should be handled carefully because of the harm their abuses cause to individuals, families and communities and therefore should not be encouraged through investments. A variation on this theme is a 'refusal to profit from' position that holds that it is ethically inappropriate to profit from, or seek to maximise the profits from, certain societally harmful or morally questionable activities such as war or the sale of addictive products.

A number of institutional investors, particularly in Europe, refuse to invest in certain activities that have been declared illegal or improper through the adoption of international conventions, treaties and standards. They argue that a broad-based consensus on conduct has been reached on these issues and that it doesn't make sense to invest in companies that engage in activities that their governments have gone on record as opposing for being in contravention to international norms and standards, some of which have the force of international law. Among these are bans on the use of landmines and anti-personnel weapons, the non-proliferation of nuclear weapons, the establishment of principles of universal human rights, conventions for fair labour practices and the establishment of international goals for protection of the environment.

For example, in 1996, the International Court of Justice delivered an Advisory Opinion that the use of, or the threat of the use of, nuclear weapons is contrary to international law.[103] In light of this ruling, manufacturers of nuclear weapons might be excluded as an industry. In addition, at a national level, certain regulations can be relevant. For example, several European countries now discourage investment in companies involved in the production of landmines and cluster bombs. In the United States a number of states and cities have passed legislation forbidding their pension funds from investing in manufacturers of tobacco products.

An additional justification for an exclusionary approach can be based on customer demand. RI products are often designed for a specific market or client base, whose expectations must be met. These expectations can spring from a variety of convictions, but in the aggregate they may have the net effect of creating a *de facto* standard for the fund. For example, in 2003 the bank-watch organisation Netwerk Vlaanderen and the peace organisations Vrede, Forum voor Vredesactie and For Mother Earth launched a campaign against the investment of Belgian banks in weapons of all kinds, creating pressure for RI funds in Belgium to limit their investments in this industry.[104] In the United States most responsible investors would expect a financial product that billed itself as socially responsible not to invest in tobacco companies.

As to the question of practicality, when it is argued that divesting from companies in whole industries leaves investors without a voice with which to advocate change, a counterargument can be made that the kinds of change investors are seeking in these cases are typically not ones that can be resolved through negotiations between investors and companies in the marketplace, but rather need regulation and governmental actions. For example, governments must deal with questions of nuclear weapons disarmament or reductions in international weapons transfers. Investors could not reasonably be expected to effect such changes through dialogue with corporations. Similarly, negotiations between investors and tobacco companies could not be expected to lead to a ban on smoking in restaurants or public places. Only government regulation can be effective here. A refusal to invest in certain industries can therefore be seen as a signal to government that market mechanisms cannot deal with the fundamental problems posed by these industries and that regulatory action is necessary.

These various non-financial justifications are not necessarily mutually exclusive. Responsible investors may avoid nuclear weapons because they believe simultaneously that these weapons are immoral and illegal and that only government can free the world of their current dangerous presence. Whatever the justification, it is important that responsible investors be transparent about the reasons for exclusion and the criteria used.

Variable factors

Responses to this case may differ depending on a number of factors, including the cultural characteristics and business strategies of the organisation, the investors' expectations, the potential to exert pressure on companies or a whole sector and the potential public policy impacts of excluding an industry.

- What are the cultural characteristics and business strategies of the organisation?
  - Is it a financial services firm comfortable with being identified with particular exclusionary standards?
  - Where and to what types of client (retail or institutional) does the organisation market its products?
  - What is the investment universe on which the financial product is based?
  - What are the potential financial impacts of the exclusionary approach at the portfolio or fund level?
- What are the general expectations of the investors to whom the product is being sold?
  - Do clients have specific demands or expectations?
  - What are the norms and values of the country or region where the fund is being sold?
  - Are there any legal or ethical requirements on particular issues that the client must adhere to?
- What is the potential of exclusionary standards to exert pressures on a company or its industry?
  - Is dialogue a reasonable means to achieve the desired changes?
  - Would rewarding and recognising the best practices in the sector make a difference?
- What is the potential for exclusionary standards to have an impact at a governmental or public policy level?

- Is there public pressure on government at the current time to bring fundamental changes to the industry?
- Would an exclusionary policy attract public attention and promote broad debate on the issue?
- Are there other responsible investors pursuing the same exclusionary approach and would the fact that multiple investors have excluded a particular industry lend weight to this approach?

Recommendations

As you approach dilemmas of this sort, you may want to:

- Understand the expectations of the firm's existing and potential clients

- Ensure that the exclusionary criteria are aligned with the purpose of the fund

- Be transparent with existing and potential clients as well as with the companies with regard to how the exclusionary criteria are defined

- Be aware of the impacts of excluding industries, both financial and societal

- Understand the potential and limits of exclusionary criteria at both the fund and the societal level

Justifying exclusions

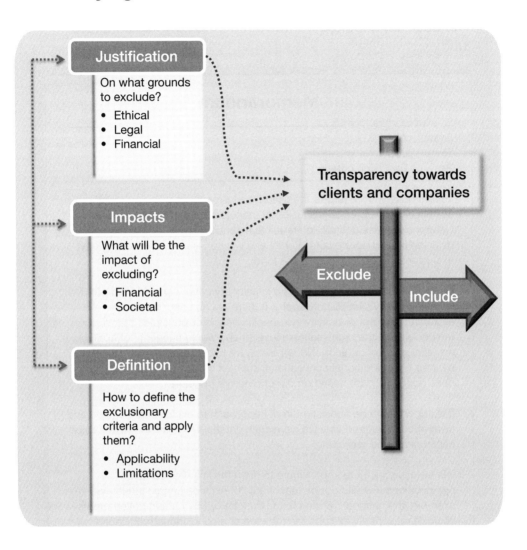

Responses from practitioners

The decision should be embedded in the principles of the fund

---

Memorandum

From: Alain
To: John
Re: Exclusion criteria

Your enquiry with regard to the use of exclusionary criteria and more specifically the potential exclusion of manufacturers of landmines and weapons of mass destruction requires extended research. The final decision should not be taken lightly and will necessitate embedding any exclusions in the principles of the fund.

Our decision to use or not to use exclusionary criteria should be grounded in our company's basic philosophy. If there are no specific reasons for excluding a sector or activity, no specific beliefs or principles that support the decision at the organisational level, then I would suggest not applying any exclusionary criterion. Nonetheless, we may always exclude what is banned by international conventions, such as cluster bombs, since that does not need to be based on our company's principles.

It is important to be consistent in our approach to exclusionary criteria and clear about what and why we have such criteria. Our decision needs to be based on strong reasoning.

It is always difficult to know where to draw the line. For certain organisations it would not be reasonable to exclude certain industries. For example, the national pension funds that are run by certain governments couldn't reasonably exclude the tobacco or alcohol industry because the government itself holds ownership in tobacco companies as well as alcohol companies. That would appear contradictory. We want to avoid contradictions.

Finally, it is crucial to ask stakeholders, our clients, or the founders of the fund what they really want, and maybe to rethink or clarify the principles of the fund in light of those responses. However, as a first step we could make international conventions a minimum requirement, meaning that if a company violates an international convention, it should be excluded.

International convention should be the threshold

Memorandum

From: George
To: Linda
Re: Exclusion criteria

I would recommend excluding the companies that manufacture landmines and weapons of mass destruction, and would draw the line there. These weapons have been forbidden through a number of global initiatives that many countries, although not all, have signed. So this is really something companies should think about as world citizens. Since often the orders come from governments, there's no way to stop the production of this sort of product – so there's no way for us to make a difference through engagement. The only option left to us is divestment.

For the other industries mentioned – uranium mining and nuclear power – it depends on what use the company puts these products to and how they manage their production. This is not like weapons of mass destruction, but depends on how the company manages the production process and its products. On the other hand, nuclear power, if managed properly, could be a good resource for energy efficiency in the future. So I'm drawing the line there. This is about whether an exclusionary policy would be useful, and about whether we have the opportunity to engage with the company to make a difference.

> ❝ In general we think that exclusionary criteria ought to be used, especially with regard to landmines and cluster bombs. Exclusionary criteria make it impossible for companies to be included in the portfolio and therefore they have no chance to improve. Whatever they do they will be excluded. You can't have an influence on those companies. ❞

> ❝ For our funds we have a number of exclusionary criteria and sometimes we get questions from outside. I think the weapons criterion is one that is never questioned. But we also exclude nuclear power and genetic engineering in agriculture and those two criteria are more subjective, and opinions on those criteria differ more from market to market. ❞

The client is the one that needs to make the decision

Memorandum

From: Mary
To: Jan
Re: Exclusion criteria

For our separately managed accounts, I would recommend letting our clients draw the line themselves, because generally our clients' objectives are twofold: (1) to invest in companies that are better aligned with their values, and (2) to use their investment leverage to help strengthen corporate social responsibility on the issues they care about.

On the other hand, for our mutual fund products, I think we need to make the decision. My position would be to be more inclusive of the kinds of concern that potential clients might have. This, I believe, will not harm the investment performance. I therefore recommend ruling out landmine manufacturers and also companies 'significantly engaged' in weapons of mass destruction.

The next question is how to define the companies' involvement in those basic business areas. My suggestion is the following:
1.   What is the market share of the company? How important is the company in this specific market?
2.   How dependent is the company on this specific activity?

From experience, we know that a 'zero tolerance' policy is generally not the best approach from a practical or legal perspective, and it does not necessarily reflect the goals of our clients. They are not ideologues and would be comfortable with a general policy that didn't try to draw too fine a line.

In the case of weapons of mass destruction, we can plot involvement out on a spectrum and see who the market leaders are or make a cut based on revenues. It may be more difficult with regard to uranium mining, and nuclear power is an even tougher question these days. If our objective is to appeal to most investors, it's probably smart to rule out companies in these two industries. Nonetheless, we need to consider whether companies with some investment in nuclear utilities are doing a lot in the alternative energy space and give them credit. No new commitment to nuclear power plants could be a positive criterion.

Finally, I am in favour of ruling out companies that are significantly involved in the production of tobacco. It may help us to appeal to a larger audience.

In the news

In constructing their world sustainability indexes, Dow Jones and FTSE have adopted two contrasting approaches when it comes to excluding certain industries. As of May 2010, the Dow Jones Sustainability Indexes provided a primary index that applied a best-in-class approach to all industries, without excluding any, although it allows investors to create their own indexes that will exclude problematic industries. FTSE's family of indexes, by contrast, excludes a limited set of industries from its indexes.

The Dow Jones Sustainability Indexes are maintained on a best-in-class approach. Working with SAM Group, Dow Jones researches the sustainability records of all companies in all industries, scores them and ranks them by their sustainability score. It then constructs its indexes by including those companies whose scores on environmental, societal and economic criteria place them in the top 10% of their industry and excluding the rest. Its basic indexes at the global, regional and national levels include all industries, but it also offers clients the option of creating special indexes that exclude companies in such industries as alcohol, tobacco, gambling, firearms and adult entertainment.[105]

The FTSE4Good family of indexes excludes companies with interests in tobacco, nuclear power stations, weapons systems and nuclear weapons systems from its family of indexes. It also excludes companies in other industries case by case if they fail to meet the indexes' criteria in such areas as environmental sustainability, stakeholder relations, human rights, supply-chain labour standards and anti-bribery efforts. FTSE4Good applies those exclusions to all its index products.[106]

Cases for comparison

Compare and contrast this case with Case 3 'Influence through voice and exit', Case 4 'Societal returns versus financial returns', Case 5 'Alleged versus confirmed illegal activity' and Case 12 'Privatisation of public services'.

Case 11
Emerging issues

- **Should responsible investors exercise caution regarding emerging technologies?**
- **How can responsible investors assess the science of emerging technologies?**
- **What should responsible investors do in cases of uncertainty about emerging technologies?**

The case

You are an RI money manager. Among your clients is an environmental foundation. For this foundation, you do not invest in oil companies because of its concern about climate change. You also do not invest in agribusiness companies that are major promoters of genetically modified organisms (GMOs). The head of the foundation, Mrs Green, asks for a meeting with you to discuss the concerns of the foundation about the emerging field of nanotechnology. She believes that the introduction of nanotechnologies into business has the potential to cause major unforeseen disasters. Scientists are now developing active chemical molecules so small that they can penetrate virtually anything and once released into the environment could have unpredictable implications for human health and other life forms. She wants your opinion about how to handle this emerging issue. You do some research on nanotechnologies and discover that the scientific community is divided on the issue. These technologies have broad applications across many industries and have the potential to improve many products and manufacturing processes. They also involve the creation of entirely new microscopic materials. Scientists don't know how these materials might act once introduced into the environment, but some believe they could cause unforeseen environmental or human health disasters.

What will be your approach in your meeting with the head of this foundation . . . ?

Dilemma for the responsible investor

This case highlights dilemmas that arise when responsible investors must confront the implications of emerging, controversial technologies, particularly those that promise substantial short-term financial benefits.

Speculation about the future implications of emerging technologies is a fundamental part of mainstream financial analysis. Industry analysts frequently speculate about whether a new software program or mobile phone innovation will lead to the next blockbuster product, whether a pharmaceutical company's new drug will revolutionise the treatment of this or that disease, whether government will subsidise the development of new alternative energy technologies, and similar questions. Their speculation is often limited to the financial implications of these new technologies for a particular company or industry and does not include their broader societal or environmental implications.

Responsible investors also assess the financial potential of new technologies, but in addition are concerned with their broader societal and environmental implications,[107] particularly as they extend to the technology's potential long-term costs to society.

For example, responsible investors might view sceptically emerging carbon capture and storage technologies, which could benefit coal-dependent electric utilities in the short run but also could delay the replacement of coal as a source of energy for generating electricity and therefore have long-term detrimental environmental implications. Or conversely, they might view positively the health benefits of preventative medicines, although these might reduce healthcare costs and thereby hurt pharmaceutical companies' profits.

The case in perspective

This case focuses on the challenges faced by responsible investors when confronted with emerging technologies. It highlights the importance of responsible investors' willingness to speculate about the future societal and environmental implications of innovations – and particularly their potential downsides – in the face of considerable uncertainties. Mainstream investors tend to focus on the potential of innovative, untried technologies for short-term financial gain and to pay less attention to their potential for long-term societal or environmental harm.

Responsible investors tend to be cautious in these assessments and willing to consider long-term societal or environmental costs that may not immediately manifest themselves. In short, they are willing to factor into their investment decisions considerations about potential negative externalities – costs that may be imposed on society or the environment at some time in the future but that

→

Responsible investors often take a cautious approach, considering as a negative the possibility that costs may be externalised onto society and are more inclined to invest in prevention than in cures. For example, responsible investors may prefer investments in a technology that eliminates toxic chemicals in the manufacturing process to investments in a technology for cleaning them up once they have escaped into the environment.

Responsible investors acknowledge the fact that new technologies can be profitable for particular companies or industries in the short term, while having the potential to cause substantial and costly problems for society or the environment in the long run. Determining when this will and will not be the case can be difficult. Analyses often require specialised expertise and assessments of contradictory arguments in highly technical matters, and even then answers may not be easily found.

will not necessarily be borne by the corporations themselves.

This complicated equation, which assesses potential short-term benefits to corporations as well as long-term costs to society, arises for responsible investors as untried technological innovations appear. Responsible investment's careful considerations of these externalities – either positive or negative – can help allocate assets in ways that minimise costly mistakes and direct finance towards societally and environmentally beneficial outcomes.

Approaches available to the responsible investor

One of the key themes raised in this case is that of speculation in the face of uncertainty. The financial community prides itself on its ability to recognise, analyse and manage various investment risks, but emerging technologies often involve unknowns and uncertainties that cannot be immediately resolved. Responsible investors can find themselves attempting to evaluate the claims of industry – which tends to minimise the possibilities of negative consequences of new profitable technologies – versus those of government, independent scientists or environmentalists, who are frequently more critical. Faced with these contradictory claims and ambiguous facts involving technical scientific matters, responsible investors have a number of choices. They can:

- Rely on the judgement of government regulators

- Weigh for themselves the conflicting assertions of industry and its critics

- Identify a trusted independent third party and rely on its judgement

- Exercise the precautionary principle and wait for further developments

One simple approach for responsible investors can be to look to government regulators for guidance in such situations. It is essentially government's job to protect society and the environment from the adoption of harmful technologies by business. Governmental organisations often mediate between an overly aggressive industry and an overly cautious citizenry. If government has approved a technology, or is in the process of evaluating it, there is a certain logic to following government's lead. Government, it can be argued, represents a reasonable consensus opinion and one that can be trusted.

At times, however, responsible investors may find that regulators are slow to act or are, to a certain degree, subject to capture by industry. In such cases, they may decide in effect to conduct their own research – to review the most credible research coming from both advocates and critics of the technology in question. In doing so, they must become an expert in the emerging field, delving deep into the literature, familiarising themselves with the technical and scientific issues involved and trusting in their own judgement to assess the technology's possible rewards and dangers. This choice implies a decision to keep up with the most current research and debate as new research emerges. Mastering the full range of technical details requires a substantial commitment of time and resources.

An alternative approach involving a considerable, but less substantial, time commitment is for responsible investors to identify an independent third party whose opinion they respect and trust. They can then rely on this party's judgement in the matter. This third party could be a scientific or environmental research organisation that has worked on similar issues in the past and has shown itself to be independent, objective, insightful and reliable. It should be an organisation that can be trusted to criticise either industry or government, or both, when appropriate. This approach has the advantage of providing responsible investors with access to thorough analysis without having to conduct complete analyses themselves.

Finally, a responsible investor may simply choose to err on the side of caution, opting to avoid companies using the technology in question until further scientific research can satisfactorily resolve the uncertainties. This approach has the advantage of thoroughness. Still, it can be a difficult approach to implement because certain technologies can rapidly become widely adopted, making avoiding companies involved with them impractical. Moreover, scientific studies can take years to complete, during which time investors taking the cautionary approach may miss out on valuable investment opportunities.

Because each new controversial technology involves its own unique set of challenges, these situations must necessarily be evaluated case by case.

Variable factors

Responses to this and similar cases may differ depending on a number of factors, such as the internal research capability of the investor, the relative potential upsides and downsides of the new technology, the degree of uncertainty in the current debate and the institutional structure of the investors and their products.

- What are the research capabilities and opportunities available to the investor?
    - Does the investor have the time and resources to evaluate the full range of studies relating to the technology?
    - Can the investor identify a third party who has already done credible analyses of the issues at stake?

- What are the potential future rewards and risks for use of the technology?
    - Are the potential downsides of the technology greater than its upsides? Or vice versa?
    - Is the uncertainty about the future implications of the technology greater for either its potential rewards or its potential risks?
    - Are the potential risks or rewards likely to be confined to a single industry or to arise across a broad number of industries and companies?

- What is the state of scientific knowledge and debate about the technology?
  - Has a relatively clear scientific consensus emerged about the technology?
  - Are there diametrically opposed findings and concerns among reputable scientists?
  - Is the trend in scientific research headed in a particular direction?
  - Are major studies underway that are likely to produce definitive results in the near future?
- What is the investment structure within which the responsible investor is operating?
  - Can the investor offer to tailor the selection of securities to the exact wishes of the client?
  - Does the investor offer a mutual fund or other pooled vehicle that necessitates the taking of a public position on the technology?

Recommendations

Responsible investors faced with similar dilemmas may want to:

- Be certain to know the basics of the current state of knowledge about the new technology in question and the arguments made for and against it

- Consult with others including experts in the field, peers in the responsible investment world and clients

- Follow the most recent developments in the field

- Be prepared to change their minds as new facts or arguments emerge

Assessing emerging issues

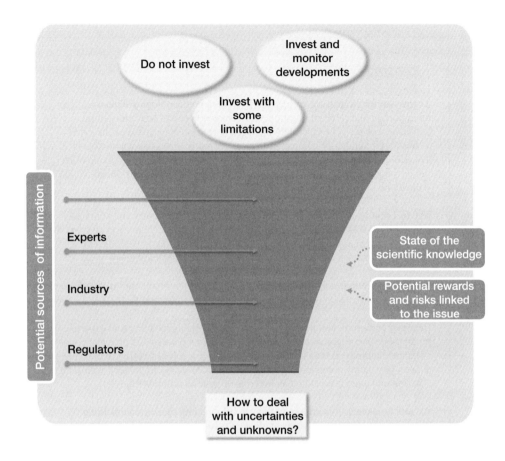

Responses from practitioners

Take a precautionary approach

Dear Mrs Green,

Thank you for expressing your concern about nanotechnologies. This is an important and relevant question. Our decision here can have substantial consequences for your separately managed account – namely the exclusion of a substantial number of companies that are currently using nanotechnologies. Therefore it cannot be considered lightly and requires considering among other things the market capitalisation of those companies potentially excluded, which will be one of the driving factors in the decision-making.

The fact that your foundation already screens out oil companies and companies with GMO involvement suggests that you are comfortable with the exclusion of whole groups of companies. However, are you now willing to exclude a third group of companies, namely the ones involved in nanotechnologies?

Our recommendation will probably be to take a precautionary approach and avoid companies that most actively use or promote nanotechnology.

Being an environmental foundation, you most probably have internal expertise on the issue and are probably better informed than we are on the pros and cons of nanotechnology. But as your money manager our role is to bring to the table data on how many companies would be excluded were we to proceed with such a criterion and what it would mean in terms of your overall portfolio.

As an RI manager, it is important to us to be a leading thinker on emerging issues, and to be able to provide guidance to the RI community as a whole from both a risk and a values perspective. We should be forward looking, but with the understanding of our own limitations and that a precautionary principle – that is, avoiding entirely companies involved in this emerging issue – may not be fully implementable.

Ultimately, the final decision is yours. We can only give you guidance on the appropriate facts and figures, and how we can best formulate criteria in ways that will meet both your financial and environmental needs.

Sincerely,

Your Money Manager

Finding the right balance

Dear Mrs Green,

We fully understand your concern with regard to nanotechnology. Although we are currently working on this issue, we cannot yet tell you at this time what our final position will be. As soon as we have reached our final decision we will inform you.

Nanotechnology is a complex issue, similar to the GMO issue. Our decision will be informed and shaped by:
- An advisory board of experts
- Lengthy discussions among people who are both for and against nanotechnology.

Throughout this process we need to strike a balance between how strictly we can impose criteria on this topic and how we can select a portfolio that will outperform others. We always maintain our investment perspective and seek to keep an adequate number of investment opportunities available for the fund. On the one hand, we are talking about defining your own position and on the other hand, we want to identify the clear leaders on the issue. If what we are demanding cannot be achieved by the leaders, we are probably demanding too much and being unrealistic in the setting of our criteria. So the case involves defining and developing the criteria and checking how realistic they are – that is the balance we are seeking.

One of the problems with this issue and others such as GMOs is that they are complex. It is difficult for many people to understand them. We are now working on a more extended description on how we apply standards to GMOs and nanotechnologies, which we will put on our website. We also have position papers on how we apply other criteria.

In the case of GMOs, all companies involved are not automatically excluded but we are very strict. We have a hierarchy of increasingly strict criteria starting with the use of genetically modified organisms in food, then the development of GMO crops and microorganisms, and finally their use in genetic engineering and the altering of human nature. The deeper the involvement, the stricter we are.

One problem is that some companies may not be transparent on the issue. So the absence of transparency itself can become an issue. For example, when we are evaluating the pharmaceutical industry, most companies are excluded for a combination of reasons, including their practices on access to medicine, patent issues, etc. Lack of transparency on GMOs and genetic engineering can also be a determining factor.

There are usually one or two companies in a sector that are proactive leaders and we tend to work with these front runners. In the pharmaceutical industry right now we have only one company we feel is a true leader on various issues.

One of the challenges with GMOs, and we expect the same with nanotechnology, is that every day there are new regulations or scientific studies that could alter our opinion or yours. This is one of the reasons why we take the ultimate decision upon ourselves.

We thank you for your trust.

Sincerely,

Your Money Manager

66 RI has a unique role here. We are often the 'canary in the coal mine' and should bring these issues to the attention of the investment community and of broader communities. 99

66 I think this is similar to some of the new technologies like carbon sequestration – no one knows if it will work or be a disaster. A lot of people consider these issues and invest in a short-term way, but we always consider the potential long-term negative impacts. 99

66 We don't want clients to just wait until everybody decides – we should be proactive. 99

- page 2 of 2 -

In the news

Nanotechnology is much in the scientific, business and popular press because of the many uncertainties arising from its pioneering use of engineering on the molecular level, a kind of engineering never before employed. In addition, its potential applications are extremely broad. It has applications in the fields of medicine, electronics, energy, cosmetics, consumer products and agriculture, among others. According to its proponents, such as the UK-based Institute for Nanotechnology[108] and the US National Nanotechnology Initiative,[109] nanotechnology has the potential to provide the solutions to long-standing medical, societal and environmental problems on a global scale.

At the current time, the health and environmental implications of nanotechnology are still not fully understood. A number of studies have raised questions about the safety of nanotechnology, and a number of advocacy groups such as Friends of the Earth and the Center for Responsible Nanotechnology have argued for additional research, a slower pace of commercial application and carefully crafted regulation.[110] Regulators worldwide have been moving slowly as they weigh the potential benefits of nanotechnology against the potential health and environmental risks of allowing the field to develop freely.

Cases for comparison

Compare and contrast this case with Case 5 'Alleged versus confirmed illegal activity' and Case 9 'Incomplete societal and environmental data'.

Case 12
Privatisation of public services

- Should a responsible investor make judgements about which goods are private and should be provided by markets and which goods are public and should be provided by governments?

- What is the relationship between responsible investment and public policy?

- To what degree does responsible investment involve taking political positions?

The case

You are an RI manager considering investing in a recently privatised water utility company. Among other things, this company owns and operates municipal water services around the world. You are enthusiastic about this investment because you believe that providing water is an important societal service and that RI clients should feel positively about a company that provides such a socially useful product. You are aware that some other water utility companies have been involved in controversies about the quality and price of the services they provide, particularly in developing countries. But the company you are recommending hasn't been involved in such controversies and you believe it is being well run. One of your clients, Mr Martin, is very unhappy about this investment. He argues that access to drinking water is a human right. Water should be free – or as nearly so as possible – and that allowing for-profit companies to market water will inevitably lead to problems and abuses. He has read reports from Latin America about privatised water companies that have charged unaffordable rates to the poor and then cut off their water supply when they could not pay. Water should be provided only by governments, and as cheaply as possible. Not only does he not want this company in his own portfolio, but he also can't understand how you can consider putting it in any of your clients' portfolios. He threatens to take his account to another RI manager if you invest in this firm at all.

What would you do . . . ?

Dilemma for the responsible investor

This case highlights an interrelated series of dilemmas that responsible investors face in relation to the privatisation of state-owned or highly regulated businesses and to the relative roles of business and government in providing services to society. For water – as well as for other services such as healthcare, roads, public transportation, infrastructure, education, national and local security, and prisons – debates rage as to the most appropriate balance between the public and private spheres, with different countries adopting a broad range of differing approaches.[111]

The wave of deregulation of government-controlled prices and the privatisation of many state-owned industries that swept the globe in the 1980s and 1990s highlighted the challenges that societies around the world face in finding the appropriate balance between government and business in the economy. During that era, a major shift from government control to market control took place in many sectors. Among those industries deregulated or privatised during that period were airlines, railways, road haulage, telecommunications, financial services, electric utilities, natural gas utilities, tobacco, defence and postal services. For certain industries – such as telecommunications – benefits flowing from privatisation and deregulation have been widely recognised. For others – such as financial services – many in and out of government have pointed to the dangers of this deregulation.

The mainstream financial community generally applauds privatisation and deregulation. By contrast, many in non-profit organisations, academia and government have called attention to costs of the recent pull-backs in oversight and control of these industries. The financial community's enthusiasm for deregulation and privatisation is reinforced by contemporary theories of finance that view the broadest possible opportunity for diversification of investment options as a positive.

The case in perspective

This case addresses the complexities that arise when responsible investors evaluate the proper balance between government and private enterprise in the provision of goods and services. The financial and business communities typically favour deregulation and privatisation on the grounds that markets are more efficient than government in setting prices and allocating resources. Responsible investors tend to factor in the benefits of public goods – for example, infrastructure, pure research, education and so on – that government, properly managed, is typically responsible for.

Weighing the benefits of market efficiencies against those of government-created public goods is never easy.

→

The dilemma for responsible investors is that, while desiring diversification options, they often are simultaneously sceptical about the role of unfettered markets and are receptive to arguments that certain services are most appropriately provided by government. Many believe that a dominant role for government may be appropriate because certain goods or services are a basic right (e.g. access to healthcare), should not be subject to the discipline of the for-profit motive (e.g. education) or should serve a utility function in society (e.g. electricity generation). For this reason, responsible investors on occasion must make independent judgements about what the appropriate balance between government and markets should be for various industries.

In practice, the answer to the question of how best to provide such services varies considerably from country to country, from culture to culture, from one era to another, from one stage of economic development to another and from industry to industry. No single formula can be applied to all circumstances. In addition, judgements about where and when business is more effective in providing services than government, or vice versa, are often highly political, involve decisions about the nature of fairness and justice in the design of a society and are subject to the local dynamics of power at any given time.

Society is not well served when government fails to deliver promised goods and services in a high-quality, efficient manner or abuses its privileged position through corruption or injustice. Similarly, society is not well served when private enterprise disregards societal and environmental welfare in a mindless pursuit of profit. It is often not clear which of the two parties, or what combination of the two, can best do the job.

Because of technological innovations, shifts in political power and uncertainties about natural resources and the environment, among other things, considerations about the appropriate roles of the public and the private will often be a challenge for responsible investors. Investors' considered attention in these matters can help societies achieve an appropriate balance between government and private enterprise when providing goods and services to the public.

Approaches available to the responsible investor

One of the key themes raised in this case is the appropriate roles of business and government in providing basic goods and services in a just and sustainable society. In confronting this issue, responsible investors can find themselves drawn into complex controversies, fraught with uncertainty and forcing them to take what are essentially political positions. The question of how to incorporate political positions into investment practice without allowing personal politics to prevail can pose a substantial dilemma. This challenge can be addressed if responsible investors:

- Acknowledge openly the political implications of many investment decisions

- Identify a generally recognised theory of just and fair action to guide their decision-making in such situations

- Give reasonable consideration to the full range of opinions available in the situation in question

- Make consistent decisions in analogous situations

Because questions of state versus private ownership are highly controversial, confronting them necessitates taking what amounts to political stands. Investors acting as fiduciaries are under an obligation not to promote their own personal interests through investments made with other people's money – that is, are obliged not to use their position for personal gain or profit.

From time to time, however, certain investment decisions involve a political debate that cannot be avoided. In such cases responsible investors should, first and foremost, be fully aware of the political context and implications of the situation and the political implications implicit in their investment decisions.

In order to form specific judgements in these situations, it is helpful to have in mind a general theory of how public and private goods and services should appropriately and fairly be provided. Many well-established theories of justice and the creation of just and sustainable societies exist – and various of these have been enshrined in internationally accepted norms and standards that have been developed and widely endorsed, notably through the auspices of the United Nations. The United Nations Universal Declaration of Human Rights is one such example.[112]

While reasonable parties may disagree on the particulars of such theories and norms, identifying a framework of widely accepted principles of fairness and justice can be useful to the responsible investor because it can guide consistent decision-making that transcends the purely personal.

Although such a framework can help generally, specific situations are often complicated and involve highly polarised positions and interpretations of facts. Responsible investors must give careful consideration to the particulars of such situations. They need to consider the positions of both sides and to take into consideration the analyses of reputable independent third parties.

Finally, it is useful for responsible investors to document their decisions to ensure that future decisions in analogous situations are reasonably consistent.

Variable factors

Responses to this case may vary depending on a number of factors, such as whether the investors are acting on their own behalf or as fiduciaries, the relative capabilities of government and industry to deliver the goods or service in question and the records of success or failure of governments and industries in similar situations. Asking questions such as the following can be helpful.

- Is the investor acting privately or as a fiduciary for others?
  - If the investors are private individuals, have they articulated a general philosophy concerning the kinds of service that government should provide, as opposed to industry?
  - If the investors are acting as fiduciaries, have they considered appropriate generally accepted principles of justice or fairness within which the provision of such services can be evaluated?

- What are the relative capabilities of the specific governments and industries to provide the services in question competently, broadly and at a fair price?
  - Does the government have a strong record of providing such services?

- Does industry have a proven record of providing such services competently and fairly in the place of government?
- Does the government have the capability to oversee and regulate the industry that would be providing this service?
- Would an industry providing the service in question undercut the government's ability to deliver the service at a later date?

• What have been the successes or failures elsewhere in similar situations?
- What lessons can be learned from other governments' conduct in other regions of the world?
- What lessons can be learned from the industry's conduct in other regions of the world?

Recommendations

As you approach dilemmas of this sort, you may want to:

• Develop a basic position on the proper balance between government and private markets when it comes to specific goods and services

• Understand that differences in national cultures, histories and regulatory environments, as well as stages of economic development, can call for differing relative roles of government and business

• Not allow personal politics to dominate your decision-making

• Grant a legitimate place in decision-making to national or international political concerns, where reasonable consensus has been established

The private and public balance

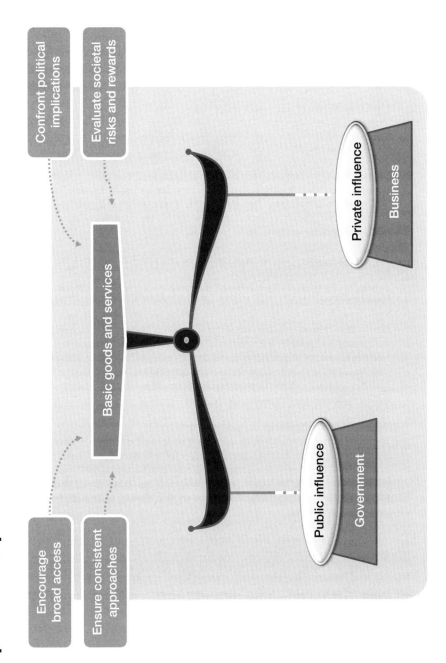

Responses from practitioners

First enter into dialogue and exclude only if necessary

Dear Mr Martin,

Water is a particularly controversial issue and we have invested in several water companies in our ethical funds.

In general, our outside advisory committee looks for the positive in investment opportunities. Their feeling is that access to water, in terms of people's health and other considerations is a social positive. Therefore when evaluating water companies our primary concerns are assuring that the positives of access and reasonable pricing are provided.

For most companies, ethical issues will arise and this is the one that you have to bear in mind when evaluating water companies. The issue you raised is widely covered in the press. We are aware of the company's activities in this area and need to be particularly transparent about our evaluation of these activities.

We do not automatically exclude water companies that have encountered problems with access and pricing. We first favour dialogue. If a water company is likely to lose a big contract in a developing country because it has gone in without really understanding the sensitivities about water distribution, we will engage in dialogue with that company about the necessity for community communication and for favourable pricing mechanisms for the poor.

We tend not to engage with the government on ethical issues, but we do engage with government when we think there have been market failures. Climate change is an example where there has been a market failure. We think climate change is a problem, and therefore the markets also need to be able to take into account climate change. But they can only do so if government takes strong steps to direct these markets. Once government has done so, that enables us to make long-term investment decisions.

We think there's a strong role for the investment community in engagement with governments because there are many examples of market failure, and market failures can have negative environmental and societal consequences, which in turn affect our investments. Government has to be responsible for correcting market failures. Theoretically, markets may correct themselves, but that can take decades. Government action can correct such failures within a year.

When we engage with governments, sometimes we'll just do it by ourselves. Sometimes we'll join with other investors. For example, the damage that

- page 1 of 2 -

corruption can do to investments is not fully understood. If government does not take a strong stand, corruption will not be punished. In that case, it's easy for the market to treat it as if it didn't matter. But corruption can harm our investments. We think that you need a very strong stand from government on anti-corruption. Coming back to the water companies: they are making a profit, but they are also increasing the availability of clean water. So we do not necessarily and automatically think of the water market as a failure. We only exclude water companies with a poor environmental track record and a pattern of fines. The key issue with water is access. Whether we are talking about a privatised company or a state-owned company makes little difference.

We hope this information helps answer your questions and concerns. Please do not hesitate to contact us again if we can be of further assistance.

Sincerely,

Your Money Manager

66 I have sympathy with the concept of public goods, but what often happens is that problems arise at the governmental level – for example, when government cannot provide access to drinkable water. What should we do in practice? Wait for the government to acquire the means to provide drinkable water or should we trust the private sector to do it? 99

66 It's a challenging question and it points out one of the limitations of RI. When we get into public policy questions and questions around decisions that governments make, RI has a limited role to play. 99

66 One industry that is privatised in some countries, but not in others, is highways. It's an issue of concern if low-income people cannot afford access to these highways or enjoy the ability to travel easily from one place to another. 99

Push the company and become engaged

Dear Mr Martin,

The issue you are raising may be realistic in other contexts, but is unrealistic in the context in which we are functioning as money managers. To begin with, water in our country is still controlled by the government. In addition, even if our government did allow privatisation, it can easily take control of the industry again after a few years if problems arise, and in the meantime it will also be in a position to exercise great influence over the industry. If a privatised company were to raise prices too high, then the government would probably step in and do something.

In our country, BOT (build, operate and transfer) is a public–private partnership scheme that is quite common for some large public service projects like highways and waste-water treatment plants. With highways, for example, the government has granted companies licences to build and operate highways for 20–30 years. The company must invest to build and operate the highway, but they can charge those who use it during that time, recoup their investment, and make a profit. After the contract expires, the government can then take over the highway and ideally make it free for the public

There have been some negative experiences with this model. Some highways were supposed to be licensed for 20–30 years, but after the licence expired the operator still charged fees, and the government did not follow up. But basically this is one way for the government to raise money to finance public services. I think it has proven successful in helping our country improve its infrastructure. Private investment in recent decades has been quite extensive in highway development, waste-water treatment and waste management. Even here, the government negotiates and controls price, and the companies need to abide by the price set by the government.

In short, we would invest in this water company, but as responsible investors we ought to push the company to communicate and become engaged with government and other stakeholders if it wants to make the case that its water prices are not unreasonable. Otherwise we do not think the company can operate in a sustainable way.

We hope this answers your questions. If you need more assistance, please do not hesitate to contact us.

Sincerely,

Your Money Manager

In the news

Privatisation of water has been of particular concern, with public protests in Bolivia, Ghana, and Uruguay and laws being passed in various localities around the world banning the privatisation of public water supplies.

Worldwide, more than 90% of water services are owned and operated by public companies. There are ten major companies whose business it is to provide fresh water services for profit. Three of those – Veolia Environnement (France), Suez Environnement (France) and RWE-AG (Germany) – are among the largest companies in the world. These companies have invested heavily in water utilities worldwide in developed countries (e.g. France, Spain and the United Kingdom), as well as in developing nations, sometimes in conjunction with the World Bank.

At the heart of the disagreement over privatisation of water services is the question of whether access to water is a human right, and a fear that ownership by private, profit-seeking companies will result in price-gouging and other business practices not aligned with the public interest. Proponents of privatisation argue that it can introduce sound business practices that can improve the efficiency, provision and quality of services, as well as lower price, in regions of the world where mismanagement by inefficient or corrupt governments is rife. In the 1990s, the World Bank and the International Monetary Fund were particularly strong supporters of the privatisation of water and sanitation services in developing countries, going so far as to make it a condition of their loans.

A case in Bolivia highlights some of the issues around water privatisation. In the late 1990s, the World Bank announced that it would not renew a $25 million loan to Bolivia, a country heavily dependent on financing from the organisation, unless it privatised its water services. The Bolivian government acquiesced and signed two agreements – one with a subsidiary of France-based Suez, and another with a consortium of companies including Biwater and Bechtel. Mass demonstrations erupted when these companies hiked rates by more than 35%, leaving many residents with prohibitively high bills and no access to water. After declaring a state of emergency, the Government of Bolivia cancelled its contracts and returned water management to a state utility company. But it was only in 2006 that a legal settlement was reached between the Government of Bolivia and the companies' consortium (Aguas del Tunari). At that time both parties agreed to drop any financial claims against the other. Water prices in Cochabamba returned to their pre-2000 levels but the water challenge continued as many people remained without water.[113]

Cases for comparison

Compare and contrast this case with Case 5 'Alleged versus confirmed illegal activity' and Case 10 'Exclusion of industries'.

Conclusion

Where do we go from here?

We might well have chosen 'Responsible Investment in 12 Cases' as the title of this book. Through these 12 cases we have tried to present, examine and reflect on the concept of responsible investment. Taken together they paint a picture of many of the more important characteristics of responsible investment, highlighting its potential as well as its limits.

In certain senses, responsible investment is no different from the practices of contemporary conventional finance. Responsible investors, just like other investors, want to make a reasonable financial return. They have certain risk tolerances. They care about certain issues more than others. Some want to be involved in the investment process, while others are content to leave all the decisions to their managers.

In other senses, however, they differ from the mainstream in important regards. Responsible investors acknowledge that their investments have societal and environmental implications and are willing to act on these implications when they rise to a certain level of concern, on either the positive or the negative side. These actions may involve the buying or selling of company stock, or dialogue with the firm. Either way, this use of voice and exit is crucial to a commitment to communications with corporate management on societal and environmental issues. This communication is important because these issues may not be foremost in the daily thoughts of managers as they conduct their daily business. This willingness to invest time and energy on issues of societal and environmental importance as part of their responsibilities to their clients is one of responsible investment's most important characteristics.

Responsible investors also tend to be long term in their view of the challenges that society and corporations contend with, and they are willing to consider issues and communicate with corporations and their stakeholders over long periods of time to see if ways of coping with these challenges can be found that will benefit society in general as well as individual interested parties.

These characteristics of responsible investors are important. If adopted collectively by major portions of the investment community, they could profoundly alter today's investment practices, which tend to be short term in their perspective and more focused on the performance of individual portfolios than on the broader market or societal implications of investment decisions.

The implications of responsible investment go beyond the particular issues raised in the 12 cases presented here and touch on more fundamental issues relating to the role of investors and business in society. Responsible investment implies a world of investors with far broader thinking and more systemic views than those of many investors in the mainstream. As we are writing this book today, responsible investment as a practice has won a certain recognition from that mainstream, but it is still unclear whether it will alter in fundamental ways today's conventional investment practices.

Ultimately, responsible investment implies a redefinition of what is called 'success' in investment, from a one-dimensional, purely financial approach to a multi-dimensional approach with an appreciation of the 'jointness' of stakeholder interests and value creation over extended periods of time. In other words, responsible investment implies a paradigm change.

We believe that contemporary finance could benefit substantially from the adoption of many responsible investment strategies and techniques. Our aim here is to promote discussions among those who are interested in responsible investment, among the experts and the inexperienced, the believers and the sceptics, in order to foster a deeper understanding of these potential benefits.

Notes

An overview of responsible investment

1   S. Lydenberg, 'Long Term Investing', paper no. 5 presented at the *Summit on the Future of the Corporation*, Boston, MA, 14 November 2007.

2   J. Post, L.E. Preston and S. Sachs, *Redefining the Corporation: Stakeholder Management and Organizational Wealth* (Stanford, CA: Stanford University Press, 2002).

3   A.L. Domini, *Socially Responsible Investing: Making a Difference in Making Money* (Chicago: Dearborn Trade, 2001).

4   C. Louche and S. Lydenberg, *Socially Responsible Investment: Difference Between Europe and United States* (Working Papers 2006/22; Gent: Vlerick Leuven Gent Management School, 2006).

5   UKSIF changed its name in 2009 from the UK Social Investment Forum to UK Sustainable Investment and Finance.

6   R. Sparkes, 'Ethical Investment: Whose Ethics, Which Investment?', *Business Ethics: A European Review* 10.3 (2001): 194-205.

7   D. Vogel, *Lobbying the Corporation: Citizen Challenges to Business Authority* (New York: Basic Books, 1978).

8   World Commission on Environment and Development, *Our Common Future* (Brundtland Report; Oxford, UK: Oxford University Press, 1987).

9   F. Déjean, J.-P. Gond and B. Leca, 'Measuring the Unmeasured: An Institutional Entrepreneur's Strategy in an Emerging Industry', *Human Relations* 57.6 (2004): 741-64; C. Louche, 'Ethical Investment: Processes and Mechanisms of Institutionalisation in the Netherlands, 1990-2002', (PhD dissertation, Erasmus University, Rotterdam, 2004, publishing.eur.nl/ir/repub/asset/1430/ESM-dissertation-003.pdf), accessed 19 March 2011.

10 M. Hobbs, 'UN PRI Enlists 400 Firms', *Financial Standard* (28 July 2008; www.financialstandard.com.au/news/view/23618), accessed 15 February 2011.

11 Social Investment Forum, *2007 Report on Socially Responsible Investing Trends in the United States: Executive Summary* (Washington, DC: US Social Investment Forum, 2008, www.socialinvest.org/resources/research), accessed 14 February 2011.

12 Eurosif, *European SRI Study 2010* (Paris: Eurosif, 2010, www.eurosif.org/research/eurosif-sri-study/european-sri-study-2010), accessed 15 February 2011.

13 Bauer, R., N. Guenster, J. Derwall and K. Koedijk, 'The Economic Value of Corporate Eco-Efficiency' (2006), available at SSRN: ssrn.com/abstract=675628, accessed 15 February 2011; P. Camejo, *The SRI Advantage: Why Socially Responsible Investing Has Outperformed Financially* (Gabriola Island, British Columbia: New Society Publishers, 2002); J. Derwall, N. Guenster, R. Bauer and K. Koedijk, 'The Eco-efficiency Premium Puzzle', *Financial Analysts Journal* 61.2 (2005): 51-63; A. Edmans, 'Does the Stock Market Fully Value Intangibles? Employee Satisfaction and Equity Prices' (2010), available at SSRN: ssrn.com/abstract=985735, accessed 15 February 2011.

14 M.L. Barnett, and R.M. Salomon, 'Beyond Dichotomy: The Curvilinear Relationship between Social Responsibility and Financial Performance', *Strategic Management Journal* 27.11 (2006): 1,101-22; R. Bauer, R. Otten and A.Tourani Rad, 'Ethical Investing in Australia: Is There a Financial Penalty?', *Pacific-Basin Finance Journal* 14.1 (2006): 33-48; N. Kreander, R.H. Gray, D.M. Power and C.D. Sinclair, 'Evaluating the Performance of Ethical and Non-ethical Funds: A Matched Pair Analysis', *Journal of Business Finance & Accounting* 32.7/8 (2005): 1,465-93; L. Renneboog, J. ter Horst and C. Zhang, 'Socially Responsible Investments: Institutional Aspects, Performance, and Investor Behaviour', *Journal of Banking & Finance* 32.9 (2008): 1,723-42; A. Rudd, 'Social Responsibility and Portfolio Performance', *California Management Review* 23.4 (1981): 55-61.

15 For those interested in an extensive annotated bibliography of academic studies on this topic, see the website maintained by the Center for Responsible Business, sristudies.org, at www.sristudies.org.

16 N. Amenc, and V. Le Sourd, *The Performance of Socially Responsible Investment and Sustainable Development in France: An Update after the Financial Crisis* (Nice, France: EDHEC-Risk Institute, 2010, docs.edhec-risk.com/mrk/000000/Press/EDHEC-Risk_Position_Paper_SRI.pdf), accessed 15 February 2011; R. Bauer, J. Derwall and R. Otten, 'The Ethical Mutual Fund Performance Debate: New Evidence from Canada', *Journal of Business Ethics* 70.2 (2007): 111-24; R. Bauer, K. Koedijk and R. Otten, 'International Evidence on Ethical Mutual Fund Performance and Investment Style', *Journal of Banking & Finance* 29.7 (2005): 1751-67; K. Benson, T. Brailsford and J. Humphrey, 'Do Socially Responsible Fund Managers Really Invest Differently?', *Journal of Business Ethics* 65.4 (2006): 337-57; M. Statman, 'Socially Responsible Mutual Funds', *Financial Analysts Journal* 56.3 (2000): 30-39.

17 M. Statman, 'Socially Responsible Mutual Funds', *Financial Analysts Journal* 56.3 (2000): 30-39.

18 J.D. Margolis and J.P. Walsh, *Misery Loves Companies: Whither Social Initiatives by Business?* (Aspen Institute's Initiative for Social Innovation through Business/Harvard Business School/University of Michigan Business School, 2001).

19 F.J. Fabozzi, K.C. Ma and B.J. Oliphant, 'Sin Stock Returns', *Journal of Portfolio Management* 35.1 (2008): 82-94; H.K. Hong, and K. Marcin, 'The Price of Sin: The Effects of Social Norms on Markets', *Journal of Financial Economics* 93.1 (2009): 15-36; I. Kim, and M. Venkatachalam, *Are Sin Stocks Paying the Price for Their Accounting Sins?* (Working Paper; Durham, NC: Duke University 2006).

20 J.J. Siegel, *The Future for Investors: Why the Tried and the True Triumph over the Bold and the New* (New York: Crown Business, 2005).

21 D. Wood and B. Hoff, *Handbook on Responsible Investment across Asset Classes* (Cambridge, MA: Institute for Responsible Investment at Harvard University, 2007, hausercenter.org/iri/?page_id=6, accessed 15 February 2011); Principles for Responsible Investment, *Report on Progress 2010* (London: PRI/UNEP/UN Global Compact, 2010, www.unpri.org/files/2010_Report-on-Progress.pdf, accessed 15 February 2011).

22 S. Lydenberg, *Corporations and the Public Interest: Guiding the Invisible Hand* (San Francisco: Berrett-Koehler, 2005).

23 Social Investment Forum, *2007 Report on Socially Responsible Investing Trends in the United States* (Washington DC: US Social Investment Forum, 2008, www.socialinvest.org/resources/pubs, accessed 14 February 2011).

24 Eurosif, *European SRI Study 2010* (Paris: Eurosif, 2010, www.eurosif.org/research/eurosif-sri-study/european-sri-study-2010, accessed 15 February 2011).

25 C.J. Cowton, 'Playing by the Rules: Ethical Criteria at an Ethical Investment Fund', *Business Ethics: A European Review* 8.1 (1999): 60-69.

26 Eurosif, *European SRI Study 2008* (Paris: Eurosif, 2008, www.eurosif.org/research/eurosif-sri-study/european-sri-study-2008, accessed on 15 February 2011); Eurosif, *European SRI Study 2010* (Paris: Eurosif, 2010, www.eurosif.org/research/eurosif-sri-study/european-sri-study-2010, accessed 15 February 2011).

27 Eurosif, *European SRI Study 2010* (Paris: Eurosif, 2010, www.eurosif.org/research/eurosif-sri-study/european-sri-study-2010, accessed 15 February 2011).

28 Principles for Responsible Investment, *Report on Progress 2010* (London: PRI/UNEP/UN Global Compact, 2010, www.unpri.org/files/2010_Report-on-Progress.pdf, accessed 15 February 2011).

29 Forum for the Future, *Sustainability Pays* (Manchester, UK: CIS, 2002, www.forumforthefuture.org/library/sustainability-pays+, accessed 15 February 2011).

30 Eurosif, *European SRI Study 2010* (Paris: Eurosif, 2010, www.eurosif.org/research/eurosif-sri-study/european-sri-study-2010, accessed 15 February 2011); US SIF, *2007 Report on Socially Responsible Investing Trends in the*

United States, 2007 (Washington DC: US Social Investment Forum, 2008,www. socialinvest.org/about/contact.cfm, accessed 14 February 2011).

31 C. Louche and S. Lydenberg, *Socially Responsible Investment: Difference Between Europe and United States* (Working Papers 2006/22; Gent: Vlerick Leuven Gent Management School, 2006).

32 US SIF, *2007 Report on Socially Responsible Investing Trends in the United States: Executive Summary* (Washington, DC: US Social Investment Forum, 2008, www.socialinvest.org/resources/research, accessed 14 February 2011).

33 Vigeo Italia, *Green, Social and Ethical Funds in Europe: 2007 Review* (Milan: Vigeo, 2007, www.vigeo.com/csr-rating-agency/en/news/news-vigeo/green-social-%26-ethical-funds-in-europe-%3A-sortie-du-rapport-2007.html, accessed 19 March 2011).

34 Japan Environment Ministry, 'SRI: An International Comparison of Investor Views', (2003), www.asria.org/publications/lib/japan/Japan_MoE.pdf, accessed 22 February 2011; K. Sakuma, and C. Louche, 'Socially Responsible Investment in Japan: Its Mechanism and Drivers', *Journal of Business Ethics* 82.2 (2008): 425-48; A. Solomon, J. Solomon and M. Suto, 'Can the UK Experience Provide Lessons for the Evolution of SRI in Japan?', *Corporate Governance: An International Review* 12.4 (2004): 552-66.

35 See Novethic, 'The French SRI Market in 2009', 2010, www.novethic.com/novethic/v3_uk/upload/Highlight%20_French_SRI_Market_2009.pdf, accessed 15 February 2011.

36 L. Albareda and F.M.R. Balaguer, 'The Challenges of Socially Responsible Investment among Institutional Investors: Exploring the Links Between Corporate Pension Funds and Corporate Governance', *Business & Society Review* 114.1 (2009): 31-57.

37 R. Massie, *Loosing the Bonds* (New York, Nan A. Talese, 1997).

38 From the 1970s through 1994 many pension funds and other institutional investors, particularly in the United States, had adopted South African divestment policies, but in general they withdrew from the RI field with the dismantling of apartheid in 1994.

39 See Eurosif website at www.eurosif.org, accessed 15 February 2011.

40 See Norges Bank Investment Management website at www.norges-bank.no, accessed 15 February 2011.

41 See Nyree Stewart, 'Swedish Buffer Funds Exclude Cluster Bomb Investment', *IPE Intelligence on European Pensions and Institutional Investment* (15 September 2008; www.ipe.com/news/Swedish_buffer_funds_exclude_cluster_bomb_investment_29115.php), accessed 15 February 2011.

42 See FFR website at www.fondsdereserve.fr, accessed 15 February 2011.

43 See PRI website at www.unpri.org, 15 February 2011.

44 See F&C Investments website at www.fandc.com/new/aboutus/Default.aspx?id=82810, 15 February 2011.

45 Hermes Investment website www.hermes.co.uk.

46 See Co-operative Bank website at www.co-operativeinvestments. co.uk/servlet/Satellite/1204616032483,CFSweb/Page/ Investments-UnitTrustsAndISAs, accessed 15 February 2011.

47 O. Wyman, 'Islamic assets to reach $1,600bn with revenues of $120bn by 2012 despite short term market volatility', 6 April 2009, www.ameinfo.com/191563. html, accessed 22 February 2011.

48 ORSE/ADEME, *Guide des organismes d'analyse sociale et environnementale* (Paris: ADEME/ORSE, 2007).

49 See F&C Investments website at www.fandc.com/new/institutional/Default. aspx?ID=80961, accessed 15 February 2011.

50 See the PRI website at www.unpri.org/workstreams/#1, accessed 15 February 2011.

51 See the GES Investment Services website at www.ges-invest.com/ pages/?ID=70, accessed 15 February 2011.

52 See the Eurosif website at www.eurosif.org/network/sifs, accessed 15 February 2011.

Case 1

53 Further reading on responsible investors' behaviour and attitudes:

Anand, P., and C.J. Cowton, 'The Ethical Investor: Exploring Dimensions of Investment Behaviour', *Journal of Economic Psychology* 14 (1992): 377-85.

Cox, P., S.B. Brammer and A. Millington, 'An Empirical Examination of Institutional Investor Preferences for Corporate Social Performance', *Journal of Business Ethics* 52.1 (2004): 27-43.

Glac, K. (2009) 'Understanding Socially Responsible Investing: The Effect of Decision Frames and Trade-off Options', *Journal of Business Ethics* 87, Supplement 1 (2009): 41-55.

Keller, C., and M. Siegrist, 'Money Attitude Typology and Stock Investment', *Journal of Behavioral Finance* 7.2 (2006): 88-96.

Lewis, A., 'A Focus Group Study of the Motivation to Invest: "Ethical/Green" and "Ordinary" Investors Compared', *Journal of Socio-Economics* 30.4 (2001): 331-42.

—— and C. Mackenzie, 'Morals, Money, Ethical Investing and Economic Psychology', *Human Relations* 53.2 (2000): 179-91.

Mackenzie, C., and A. Lewis, 'Morals and Markets: The Case of Ethical Investing', *Business Ethics Quarterly* 9.3 (1999): 439-52.

McLachlan, J., and J. Gardner, 'A Comparison of Socially Responsible and Conventional Investors', *Journal of Business Ethics* 52.1 (2004): 11-25.

Rosen, B.N., D.M. Sandler and D. Shani, 'Social Issues and Socially Responsible Investment Behavior: A Preliminary Empirical Investigation', *Journal of Consumer Affairs* 25.3 (1991): 221-34.

Statman, M., 'Quiet Conversations: The Expressive Nature of Socially Responsible Investors', *Journal of Financial Planning* 21.2 (2008): 40-46.

54 J.R. Gardner, 'The Robert Wood Johnson Foundation: 1974–2002', in S.L. Isaacs and J.R. Knickman (eds.), *The Robert Wood Johnson Foundation Anthology. Volume X. To Improve Health and Health Care* (Princeton, NJ: Robert Wood Johnson Foundation, 2006, www.rwjf.org/files/publications/books/2007/AnthologyX_CH09.pdf, accessed 15 February 2011).

55 Robert Wood Johnson Foundation, *SmokeLess States® National Tobacco Policy Initiative* (Princeton, NJ: Robert Wood Johnson Foundation, July 2009, www.rwjf.org/reports/npreports/smokeless.htm, accessed 15 February 2011); J.E. Cohen, M.J. Ashley, R. Ferrence, J.M. Brewster and A.O. Goldstein, 'Institutional Addiction to Tobacco', *Tobacco Control* 8 (1999): 70-74, tobaccocontrol.bmj.com/content/8/1/70.extract, accessed 15 February 2011.

56 Jessie Smith Noyes Foundation, 'Investment Policy', www.noyes.org/taxonomy/term/10, accessed 15 February 2011; Jessie Smith Noyes Foundation, 'About Us', www.noyes.org/taxonomy/term/15, accessed 15 February 2011.

57 OPERS, 'Iran and Sudan Divestment, Summary of Current Iran, Sudan Activities', https://www.opers.org/news/hb151, accessed 15 February 2011; OPERS, 'Iran and Sudan Divestment, OPERS Adopts Iran, Sudan Divestment Policy', https://www.opers.org/news/hb151/7.shtml, accessed 15 February 2011; State of Ohio General Assembly, www.legislature.state.oh.us/bills.cfm?ID=127_HB_151, accessed 15 February 2011.

58 ABP, 'Environmental, Social and Corporate Governance (ESG)', www.abp.nl/abp/abp/english/about_abp/investments/esg, accessed 15 February 2011; ABP, 'Research and Integration of ESG Factors', www.abp.nl/abp/abp/english/about_abp/investments/esg/esg_in_practice/research_and_integration_esg_factors.asp, accessed 15 February 2011.

Case 2

59 For further reading see J. Montier, *Behavioural Investing* (Chichester, UK: John Wiley, 2007).

60 E. Blass, 'Ipod City admits labor law violations', *Engadget* (26 June 2006, www.engadget.com/2006/06/26/ipod-city-admits-labor-law-violations, accessed 15 February 2011).

61 *China CSR*, 'Apple releases report on Foxconn's China labor Issues', www.chinacsr.com/en/2006/08/21/672-apple-releases-report-on-foxconns-china-labor-issues, 21 August 2006, accessed 15 February 2011.

62 J. Chan, 'High Tech: No Rights? A One Year Follow Up Repport [sic] On Working Conditions on China's Electronic Hardware Sector', Amsterdam, Centre for Research on Multinational Corporations, somo.nl/publications-en/Publication_2569/?searchterm=, accessed 19 March 2011.

63 Y. Iwatani Kane, 'Reports of Suicide in China Linked to Missing iPhone', *Wall Street Journal* (21 July 2009, blogs.wsj.com/digits/2009/07/21/reports-of-suicide-in-china-linked-to-missing-iphone, accessed 15 February 2011); D. Barboza, 'IPhone Maker in China Is Under Fire After a Suicide',

New York Times (26 July 2009, www.nytimes.com/2009/07/27/technology/companies/27apple.html, accessed 15 February 2011); B. Hill, 'Apple Responds to Foxconn Suicides, Foxconn Website Hacked', *DailyTech* (26 May 2010, www.dailytech.com/Apple+Responds+to+Foxconn+Suicides+Foxconn+Website+Hacked/article18512.htm, accessed 15 February 2011).

64 K. Hille, 'Foxconn Raises Pay by 30% in China', *Financial Times* (3 June 2010): 14.

65 D. McDougall, 'Child Sweatshop Shame Threatens Gap's Ethical Image', *The Observer* (28 October 2007, www.guardian.co.uk/business/2007/oct/28/ethicalbusiness.india, accessed 15 February 2011); D. McDougall, 'Gap Plans "Sweatshop Free" Labels', *The Observer* (4 November 2007, www.guardian.co.uk/business/2007/nov/04/3, accessed 15 February 2011); Gap Inc. Code of Conduct, 'How we do business is as important as what we do', www.gapinc.com/content/dam/gapincsite/documents/COBC/Code_English.pdf, accessed 18 March 2011.

Case 3

66 A.O. Hirschman, *Exit, Voice, and Loyalty: Responses to Decline in Firms, Organizations, and States* (Cambridge, MA: Harvard University Press, 1970).

67 B. Baue, 'Norwegian Government Pension Fund Dumps Wal-Mart and Freeport on Ethical Exclusions', *Social Funds* (16 June 2006, www.socialfunds.com/news/article.cgi/2034.html, accessed 15 February 2011); Norway Ministry of Finance, 'Responsible Investment, Recommendation of 15 November 2005', unofficial English translation, www.regjeringen.no/en/dep/fin/Selected-topics/the-government-pension-fund/responsible-investments/Recommendations-and-Letters-from-the-Advisory-Council-on-Ethics/Recommendation-of-15-November-2005.html?id=450120, accessed 15 February 2011.

68 JCPenney, 'JCPenney's Commitment to Social Responsibility', www.jcpenney.net/about/social_resp/default.aspx, accessed 15 February 2011.

69 Urban Outfitters, Inc., www.urbanoutfitters.com/urban/help/helpinfo.jsp, accessed 15 October 2010.

Case 4

70 Further reading on the relationship between financial performance and responsible corporation/responsible investment:
Amenc, N., and V. Le Sourd, *The Performance of Socially Responsible Investment and Sustainable Development in France: An Update after the Financial Crisis* (Nice, France: EDHEC-Risk Institute, 2010, docs.edhec-risk.com/mrk/000000/Press/EDHEC-Risk_Position_Paper_SRI.pdf, accessed 19 March 2011).

Barnett, M.L., and R.M. Salomon, 'Beyond Dichotomy: The Curvilinear Relationship Between Social Responsibility and Financial Performance', *Strategic Management Journal* 27.11 (2006): 1,101-22.

Bauer, R., J. Derwall and R. Otten, 'The Ethical Mutual Fund Performance Debate: New Evidence from Canada', *Journal of Business Ethics* 70.2 (2007): 111-24.

——, K. Koedijk and R. Otten, 'International Evidence on Ethical Mutual Fund Performance and Investment Style', *Journal of Banking & Finance* 29.7 (2005): 1,751-67.

——, R. Otten and A. Tourani Rad, 'Ethical Investing in Australia: Is There a Financial Penalty?', *Pacific-Basin Finance Journal* 14.1 (2006): 33-48.

Beal, D., M. Goyen and P. Philips, 'Why Do We Invest Ethically?', *Journal of Investing* 14 (2005): 66-77.

Benson, K., T. Brailsford and J. Humphrey, 'Do Socially Responsible Fund Managers Really Invest Differently?', *Journal of Business Ethics* 65.4 (2006): 337-57.

Boutin-Dufresne, F., and P. Savaria, 'Corporate Social Responsibility and Financial Risk', *Journal of Investing* 13.1 (2004): 57-66.

Derwall, J., 'The Economic Virtues of SRI and CSR' (PhD thesis, Erasmus University, Rotterdam, 2007, publishing.eur.nl/ir/repub/asset/8986/ESP20 07101F%26A9058921328DERWALL.pdf, accessed 18 March 2011).

Diltz, J.D., 'Does Social Screening Affect Portfolio Performance?', *Journal of Investing* (Spring 1995): 64.

Geczy, C.C., R.F. Stambaugh and D. Levin, *Investing in Socially Responsible Mutual Funds* (Working Paper; Philadelphia, PA: Rodney L. White Center of Financial Research, The Wharton School, University of Pennsylvania, 2005, papers.ssrn.com/sol3/papers.cfm?abstract_id=416380, accessed 15 February 2011).

Hoepner, A.G.F., 'Portfolio Diversification and Environmental, Social or Governance Criteria: Must Responsible Investments Really be Poorly Diversified?', Working Paper, draft, March 2010, www.unpri.org/academic10/Paper_1_Andreas_Hoepner_Portfolio%20diversification%20 and%20environmental,%20social%20or%20governance%20criteria. pdf, accessed 15 February 2011.

Lee, D.D., and R.W. Faff, 'Corporate Sustainability Performance and Idiosyncratic Risk: A Global Perspective', *Financial Review* 44.2 (2009): 213-37.

McWilliams, A., and D. Siegel, 'Corporate Social Responsibility and Financial Performance: Correlation or Misspecification?', *Strategic Management Journal* 21 (2000): 603-609.

Murphy, D.L., R.E. Shrieves and S.L. Tibbs, 'Understanding the Penalties Associated with Corporate Misconduct: An Empirical Examination of Earnings and Risk', *Journal of Financial & Quantitative Analysis* 44.1 (2009): 55-83.

Orlitzky, M., 'Corporate Social Performance and Financial Performance: A Research Synthesis', in A. Crane, A. McWilliams, D. Matten, J. Moon and

D. Siegel (eds.), *The Oxford Handbook of CSR* (Oxford, UK: Oxford University Press, 2008): 113-34.

—— and J.D. Benjamin, 'Corporate Social Performance and Firm Risk: A Meta-analytic Review', *Business & Society* 40.4 (2001): 369-96.

Rudd, A., 'Social Responsibility and Portfolio Performance', *California Management Review* 23.4 (1981): 55-61.

Sethi, P.S., 'Investing in Socially Responsible Companies is a Must for Public Pension Funds: Because There is No Better Alternative', *Journal of Business Ethics* 56.2 (2005): 99-129.

Statman M., 'Socially Responsible Mutual Funds', *Financial Analysts Journal* 56.3 (2000): 30-39.

71 S. Lydenberg, 'Long Term Investing', paper no. 5 presented at the *Summit on the Future of the Corporation*, Boston, MA, 14 November 2007.

72 J. Emerson, 'The Blended Value Proposition', *California Management Review* 45 (2003): 35-51.

73 *MarketWatch*, 'DJ US Railroads Index', *MarketWatch*, www.marketwatch.com/tools/industry/indchart.asp?bcind_ind=2775&bcind_sid=171585&bcind_o_symb=&indchart.x=15&indchart.y=17&bcind_period=1yr&bcind_compidx=aaaaa%3A0&bcind_comp=&bcind_compind=3353%7E171499, accessed 15 February 2011.

74 *MarketWatch*, 'DJ US Soft Drinks Index', *MarketWatch*, www.marketwatch.com/tools/industry/indchart.asp?bcind_ind=3537&bcind_sid=171597&bcind_o_symb=&indchart.x=26&indchart.y=15&bcind_period=1yr&bcind_compidx=aaaaa%3A0&bcind_comp=&bcind_compind=7577%7E171613, accessed 15 February 2011.

75 UCLA Center for Health Policy Research, 'Bubbling Over: New Research Shows Direct Link Between Soda and Obesity', www.healthpolicy.ucla.edu/NewsReleaseDetails.aspx?id=35, 17 September 2009, accessed 15 February 2011; San Francisco Department of Public Health, Environmental Health Section, 'Bottled Water vs Tap Water: Making a Healthy Choice', www.sfphes.org/water/FactSheets/bottled_water.htm, April 2004, accessed 15 February 2011.

Case 5

76 M. Friedman, 'The Social Responsibility of Business is to Increase its Profits', *New York Times Magazine*, 13 (September 1970).

77 Court of the First Instance of the European Communities, 'Judgment of the Court of First Instance in Case T-201/04', Press release No 63/07, curia.europa.eu/en/actu/communiques/cp07/aff/cp070063en.pdf, 17 September 2007, accessed 15 February 2011.

EurActiv, 'New EU Fine Brings Microsoft Bill to 1.7 Billion', 28 February 2008, www.euractiv.com/en/infosociety/new-eu-fine-brings-microsoft-bill-17/article-170596, accessed 15 February 2011.

Liedtke, M., 'Microsoft Ruling Overturned', *Washington Post* (3 March 2005, www.washingtonpost.com/wp-dyn/articles/A2828-2005Mar2.html, accessed 15 February 2011).

Meller, P., 'Update: EU Fines Microsoft Another $1.3 Billion for Antitrust Abuse', *InfoWorld* (27 February 2008, www.infoworld.com/d/security-central/update-eu-fines-microsoft-another-13-billion-antitrust-abuse-761, accessed 15 February 2011).

Wired, 'U.S. v. Microsoft: Timeline', *Wired* (4 November 2002, www.wired.com/techbiz/it/news/2002/11/35212, accessed 15 February 2011).

78 US Department of Justice, Criminal Division, Fraud Section to Davis Polk and Wardell, 'United States v. Siemens Aktiengesellschaft', 15 December 2008, www.justice.gov/opa/documents/siemens.pdf, accessed 15 February 2011; Shearman & Sterling, 'U.S. v. Siemens Aktiengesellschaft', 2008, fcpa.shearman.com/?s=matter&mode=form&id=200, accessed 15 February 2011; C. O'Reilly and K. Matussek, 'Siemens to Pay $1.6 Billion to Settle Bribery Cases (Correct)', *Bloomberg* (16 December 2008, www.bloomberg.com/apps/news?pid=newsarchive&refer=europe&sid=aCBrlMXEIYxs, accessed 15 February 2011).

Case 6

79 Wal-Mart Stores Inc., 'Wal-Mart CEO Lee Scott Unveils "Sustainability 360"', walmartstores.com/pressroom/news/6237.aspx, 1 February 2007, accessed 15 February 2011.

80 B. Baue, 'The Latest Corporate Social Responsibility News: The Greening of Wal-Mart?', *CSRwire Weekly News Alert* (15 July 2008, www.csrwire.com/press_releases/15529-The-Latest-Corporate-Social-Responsibility-News-The-Greening-of-Wal-Mart-, accessed 15 February 2011); WalMartWatch, 'Walmart Appealing Dukes to SCOTUS', walmartwatch.com, 27 August 2010, accessed 15 February 2011; Academic dictionaries and encyclopedias, 'Criticism of Wal-Mart', en.academic.ru/dic.nsf/enwiki/523845, accessed 15 February 2011.

Case 7

81 Further reading on engagement and shareholder activism:
Clark, G.L., and T. Hebb, 'Pension Fund Corporate Engagement', *Industrial Relations* 59.1 (2004): 142-71.
——, J. Salo and T. Hebb, 'Social and Environmental Shareholder Activism in the Public Spotlight: US Corporate Annual Meetings, Campaign Strategies, and Environmental Performance, 2001–04', *Environment and Planning A* 40.6 (2008): 1,370-90.
Graves, S.B., K. Rehbein and S. Waddock, 'Fad and Fashion in Shareholder Activism: The Landscape of Shareholder Resolutions 1988–1998', *Business and Society Review* 106.4 (2001): 293-314.

Hoffman, A., 'A Strategic Response to Investor Activism', *Sloan Management Review* 37.2 (1996): 51-64.

82 S. Fukuda-Parr, N. Woods *et al.*, *Human Development Report 2002: Deepening Democracy in a Fragmented World* (New York: Oxford University Press for United Nations Development Programme, 2007, hdr.undp.org/en/media/HDR_2002_EN_Complete.pdf, accessed 20 March 2011).

83 S.A. Waddock, 'Understanding Social Partnerships: An Evolutionary Model of Partnership Organizations', *Administration & Society* 21.1 (1989): 78-100.

84 J. Andriof, 'Managing Social Risk through Stakeholder Partnership Building', (PhD thesis, Warwick University, UK, 2000).

85 M. Brune, 'Home Depot announces commitment to stop selling old growth wood; Announcement validates two-year grassroots environmental campaign', www.commondreams.org/pressreleases/august99/082699c.htm, 26 August 1999, accessed 15 February 2011; Rainforest Action Network, *One Thing Stronger, Annual Report 1999–2000* (San Francisco: Rainforest Action Network, ran.org/fileadmin/materials/executive/annual_reports/RAN_AnnualReport2000.pdf, 2000, accessed 15 February 2011).

Case 8

86 The World Value Survey is a good illustration of how values vary throughout the world. How can one have one concept of RI when norms and values differ so much? The World Values Survey is an ongoing academic project by social scientists to assess the state of sociocultural, moral, religious and political values of different cultures around the world. For more information, see the World Value Survey website: www.worldvaluessurvey.org, accessed 18 March 2011.

87 N. Dawar and P. Parker, 'Marketing Universals: Consumers' Use of Brand Name, Price, Physical Appearance, and Retailer Reputation as Signals of Product Quality', *Journal of Marketing* 58 (1994): 81-95.

88 HSBC Amanah, 'Top 500 Islamic financial institutions ranking shows Islamic finance continues double digit growth despite global crisis', www.hsbcamanah.com/amanah/newsroom/press-releases/islamic_fin.html, 5 November 2009, accessed 15 February 2011.

89 M.A. El-Gamal, *Overview of Islamic Finance* (Occasional Paper 4; US Department of the Treasury, Office of International Affairs, August 2006, www.nzibo.com/IB2/Overview.pdf, accessed 20 March 2011); E. Eaves 'God And Mammon', *Forbes.com* (21 February 2008, www.forbes.com/2008/04/21/islamic-banking-interest-islamic-finance-cx_ee_islamicfinance08_0421intro.html, accessed 15 February 2011).

90 Republic of South Africa, 'Government Gazette', 9 January 2004, www.info.gov.za/view/DownloadFileAction?id=68031, accessed 15 February 2011; BEE, 'How Does the BEE Social Program Work', www.bee.co.za/Content/Information.aspx, accessed 15 February 2011.

91 Johannesburg Stock Exchange, 'Introduction to SRI Index', www.jse.co.za/About-Us/SRI/Introduction_to_SRI_Index.aspx, accessed 15 February 2011.

Case 9

92 Further reading on communication and responsible investment:
Friedman, A.L., and S. Miles, 'Socially Responsible Investment and Corporate Social and Environmental Reporting in the UK: An Explorative Study', *British Accounting Review* 33 (2001): 523-48.
Hockerts, K., and L. Moir, 'Communicating Corporate Responsibility to Investors: The Changing Role of the Investor Relations Function', *Journal of Business Ethics*, Part 2, 52.1 (2004): 85-98.

93 KPMG, 'International Survey of Corporate Responsibility Reporting 2008', www.kpmg.com/CN/en/IssuesAndInsights/ArticlesPublications/Pages/Corporate-responsibility-survey-200810-o.aspx, accessed 15 February 2011.

94 J.A.Welsh and J.F. White, 'A Small Business Is Not a Little Big Business', *Harvard Business Review* 59.4 (1981) 18-32.

95 Global Reporting Initiative, 'Small and Medium Enterprise', www.globalreporting.org/WhoAreYou/SME, accessed 15 February 2011.

96 IBM, 'Corporate Responsibility Report 2009', www.ibm.com/ibm/responsibility, accessed 15 February 2011.

97 Dell Company, 'Corporate Responsibility 2010', content.dell.com/us/en/corp/cr.aspx, accessed 15 February 2011.

98 Logitech, 'Logitech's Corporate Governance Policies and Philosophies', ir.logitech.com/governance.cfm?cl=us,en, accessed 15 February 2011.

99 Avid, 'Corporate Governance', ir.avid.com/governance.cfm, accessed 15 February 2011.

Case 10

100 For a discussion on exclusion see S. Colle and J. York, 'Why Wine is not Glue? The Unresolved Problem of Negative Screening in Socially Responsible Investing', *Journal of Business Ethics*, Supplement 3, 84 (2009): 83-95.

101 A. Rudd, 'Social Responsibility and Portfolio Performance', *California Management Review* 23.4 (1981): 55-61.

102 For more information see the website sristudies.or, which provides a summary of many studies on the social and financial performance relationship.

103 International Court of Justice, 'Legality of the Threat or Use of Nuclear Weapons', Advisory Opinion of 8 July 1996, www.fas.org/nuke/control/icj/text/9623.htm, accessed 22 February 2011.

104 For more information see the campaign report of Netwerk Vlaanderen 'Banks Disarm(ed)', April 2005, available at www.netwerkvlaanderen.be/en/index.php?option=com_content&task=view&id=232, accessed 20 March 2011.

105 Dow Jones Sustainability Indexes, www.sustainability-index.com, accessed 15 February 2011.

106 FTSE4Good Index Series, www.ftse.com/Indices/FTSE4Good_Index_Series/index.jsp, accessed 15 February 2011.

Case 11

107 Further reading on nanotechnology:

Desmartins, J.P., and H. Palmier, *Nanotechnologies: There are Still Plenty of Opportunities and Uncertainties at the Bottom* (Paris: ODDO Securities, 2006).

Gaber, F., and C. Butler, 'Guidance for Investors on Nanotechnology', *Ethix*, July 2010.

Helland, A., and H. Kastenholz, 'Development of Nanotechnology in Light of Sustainability', *Journal of Cleaner Production* 16.8/9 (2008): 885-88.

Novethic, 'Nanotechnologies. Risques, opportunités ou tabou: quelle communication pour les entreprises européennes?', www.novethic.fr/novethic/v3/les-debats-thematiques.jsp, accessed 15 February 2011.

Responsible NanoCode, www.responsiblenanocode.org, accessed 15 February 2011.

Roco, M.C., and W.S. Bainbridge, 'Societal Implications of Nanoscience and Nanotechnology: Maximizing Human Benefit', *Journal of Nanoparticle Research* 7 (2005): 1-13.

108 Institute of Nanotechnology, www.nano.org.uk, accessed 15 February 2011.

109 National Nanotechnology Initiative, www.nano.gov, accessed 15 February 2011.

110 Friends of the Earth, 'Nanotechnology Campaign', www.foe.org/healthy-people/nanotechnology-campaign, accessed 15 February 2011; Center for Responsible Nanotechnology, www.crnano.org, accessed 15 February 2011; L. Dye, 'Big Debate Over Small Science', *ABC News* (27 February 2008, abcnews.go.com/Technology/DyeHard/story?id=4348729&page=1, accessed 15 February 2011).

Case 12

111 For more discussion on the water issue:

Corporate Accountability International, 'Thirsty for Change: Why a Shift in World Bank Practices Will Help Solve the Global Water Crisis', www.stopcorporateabuse.org/privatization, 2010, accessed 15 February 2011.

Gunatilake, H., and M.J.F. Carangal-San Jose, *Privatization Revisited: Lessons from Private Sector Participation in Water Supply and Sanitation in Developing Countries* (Economics and Research Department [ERD] Working Paper 115; Asian Development Bank, 2008).

Waterjustice.org, www.waterjustice.org, accessed 15 February 2011.

World Bank Water Supply & Sanitation, www.worldbank.org, accessed 15 February 2011.

112 www.un.org/en/documents/udhr, accessed 19 March 2011.

113 S. Sadiq, 'Timeline Cochabamba Water Revolt', *Frontline World*, www.pbs. org/frontlineworld/stories/bolivia/timeline.html, accessed 15 February 2011; *BBC News*, 'Bolivia protests claim further lives' *BBC News*, 10 April 2000, cdnedge.bbc.co.uk/1/hi/world/americas/707690.stm, accessed 15 February 2011); A. Shah, 'Water and Development', *Global Issues*, 6 June 2010, www. globalissues.org/article/601/water-and-development, accessed 15 February 2011); G. Aziakou, 'UN Declares Access to Clean Water A Human Right', www. google.com/hostednews/afp/article/ALeqM5gFw3sC1VZUGBBXghGSeA-v RwYQoA, 28 July 2010, accessed 15 February 2011.

Additional resources

For those interested in learning more about responsible investment, a number of websites provide valuable resources.

Social investment trade associations

A good starting point for general information about responsible investment is the websites of the social investment trade associations in various countries around the world.

For North America, the two principle websites are those of the United States Social Investment Forum at www.socialinvest.org and the Canadian Social Investment Organization at www.socialinvestment.ca.

In Europe, Eurosif is an umbrella organisation for the various national European social investment fora. Its website links these SIFs as well as others around the world. See www.eurosif.org.

For background on RI in Asia, the Association for Sustainable and Responsible Investment in Asia (ASrIA) is a good starting point. Its website (www.asria.org) provides, among other things, a database of the more than 400 RI funds in various countries in Asia.

In addition, the Principles for Responsible Investment, in partnership with the UNEP Financial Initiative and the United Nations Global Compact, is a coalition of institutional investors and money managers committed to the concepts of responsible investment. Its website at www.

unpri.org is an excellent starting point for background on institutional investor involvement in responsible investment.

News about responsible investment

To keep up with news about responsible investment, a number of journals and online publications are helpful. Among these are: *CSRwire* at www.csrwire.com; *Ethical Corporation* at www.ethicalcorp.com; *Ethical Performance* at www.ethicalperformance.com; *GreenMoney Journal* at www.greenmoneyjournal.com; *Novethic* at www.novethic.com; *Responsible Investor* at www.responsible-investor.com; and *SocialFunds.com* at www.socialfunds.com.

Academic publications

For those interested in academic studies on responsible investment, the websites maintained by the Sustainable Investment Research Platform at www.sirp.se, sristudies.org (Studies of Socially Responsible Investing) at www.sristudies.org, and the PRI Academic Network at academic.unpri.org provide coverage of much current and past academic work in this area.

Research centres that cover important aspects of responsible investment topics include AccountAbility (www.accountability.org), Global Investment Impact Network (www.thegiin.org), and SustainAbility (www.sustainability.com). Three academic think-tanks addressing issues related to responsible investment are the European Centre for Corporate Engagement (www.corporate-engagement.com), Elfenworks Center for the Study of Fiduciary Capitalism (www.stmarys-ca.edu/fidcap), and Initiative for Responsible Investment (hausercenter.org/iri).

Rating organisations

Organisations researching the corporate social accountability records of publicly traded corporations include EIRIS (Experts in Responsible Investment Solutions) (www.eiris.org) in the United Kingdom; Jantzi-

Sustainalytics (www.sustainalytics.com) in Canada and the Netherlands; MSCI ESG Research and Screening (www.msci.com); SAM Group (www.sam-group.com) in Switzerland; and Vigeo (www.vigeo.com) in France.

For more information on rating organisation, ORSE (Observatoire sur la Responsabilité Sociétale des Entreprises) has published a guide to sustainability organisations (2007). It provides an overview of 29 rating organisations with activities that consist of, or include, analysis and rating of the sustainability performance of companies. The guide is available at www.orse.org.

Selected books

Ambachtsheer, K., *Pension Revolution: A Solution to the Pensions Crisis* (Hoboken, NJ: Wiley Finance, 2007).

Crane, A., A. McWilliams, D. Matten, J. Moon and D.S. Siegel (eds.), *The Oxford Handbook of Corporate Social Responsibility* (Oxford, UK: Oxford University Press, 2008).

Domini, A.L., *Socially Responsible Investing: Making a Difference in Making Money* (Chicago: Dearborn Trade, 2001).

—— and P.D. Kinder, *Ethical Investing* (Reading, MA: Addison-Wesley, 1986).

Eccles, R.G., and M. Krzus, *One Report: Integrated Reporting for a Sustainable Strategy* (New York: John Wiley, 2010).

Hawley, J., and A. Williams, *The Rise of Fiduciary Capitalism: How Institutional Investors Can Make Corporate America More Democratic* (Philadelphia, PA: University of Pennsylvania Press, 2000).

Kinder, P., S.D. Lydenberg and A.L. Domini, *Investing for Good: Making Money While Being Socially Responsible* (New York: Harper Business, 1994).

Krosinsky, C., and N. Robins (eds.), *Sustainable Investing: The Art of Long-Term Performance* (London, Earthscan, 2008).

Kurtz, L., 'Socially Responsible Investment and Shareholder Activism', in A. Crane, A. McWilliams, D. Matten, J. Moon and D.S. Siegel (eds.), *The Oxford Handbook of Corporate Social Responsibility* (Oxford, UK: Oxford University Press, 2008).

Lane, M., *Profitable Socially Responsible Investing? An Institutional Investor's Guide* (New York: Institutional Investor Books, 2005).

Lydenberg, S., *Corporations and the Public Interest: Guiding the Invisible Hand* (San Francisco: Berrett-Koehler, 2005).

Sparkes, R., *Socially Responsible Investment: A Global Revolution* (Chichester, UK: John Wiley, 2002).

—— *The Ethical Investor* (London: HarperCollins, 1995).

Sullivan, R., and C. Mackenzie, *Responsible Investment* (Sheffield, UK: Greenleaf Publishing, 2006).

—— (2011) *Valuing Corporate Responsibility: How Do Investors Really Use Corporate Responsibility Information?* (Sheffield, UK: Greenleaf Publishing).

Vogel, D. (1978) *Lobbying the Corporation: Citizen Challenges to Business Authority* (New York: Basics Books).

—— (2005) *The Market for Virtue: The Potential and Limits of Corporate Social Responsibility* (Brookings Institution Press).

Selected reports

Eurosif, *European SRI Study 2010* (Paris: Eurosif, 2010, www.eurosif.org/research/eurosif-sri-study/european-sri-study-2010, accessed 15 February 2011).

Freshfields Bruckhaus Deringer, *A Legal Framework for the Integration of Environmental, Social and Governance Issues into Institutional Investment* (Geneva: UNEP Finance Initiative, 2005, www.unepfi.org/fileadmin/documents/freshfields_legal_resp_20051123.pdf, accessed 15 February 2011).

Hawken, P., *How the SRI Industry Has Failed to Respond to People Who Want to Invest With Conscience and What Can Be Done to Change It* (Sausalito, CA: Natural Capital Institute, 2004).

Higgs, C., and H. Wildsmith, *Responsible Investment Trustee Toolkit* (London: UKSIF, 2005, www.uss.co.uk/Documents/Just%20Pensions%20Responsible%20Investor%20Trustee%20Tollkit%202005.pdf, accessed 15 February 2011).

Marathon Club, *Guidance Note for Long-Term Investing* (London: Marathon Club, 2007, www.marathonclub.co.uk/Docs/MarathonClubFINALDOC.pdf, accessed 15 February 2011).

Mercer Investment Consulting, *The Language of Responsible Investment* (London: Mercer Investment Consulting, London, 2008, www.mercer.com/referencecontent.htm?idContent=1210745, accessed 15 February 2011).

ORSE and ADEME (Observatoire sur la Responsabilité Sociétale des Entreprises/Agence de l'Environnement et de la Maîtrise d'Energie), *Guide des organismes d'analyse sociale et environnementale* [Guide to Sustainability Analysis Organisations] (Paris: ADEME/ORSE, 2007).

Principles for Responsible Investment, *Report on Progress 2010* (London: PRI/UNEP/UN Global Compact, 2010, www.unpri.org/files/2010_Report-on-Progress.pdf, accessed 15 February 2011).

Responsible Investor, 'Responsible Investment Landscape 2008: Asset Owners', *Responsible Investor* (2008, www.responsible-investor.com/reports/reports_page/ri_landscape_2008_asset_owners, accessed 15 February 2011).

UN Global Compact, *Who Cares Wins: Connecting Financial Markets to a Changing World: Recommendations by the financial industry to better integrate environmental, social and governance issues in analysis, asset management and securities brokerage* (New York: UN Global Compact, 2004, www.unglobalcompact. org/docs/issues_doc/Financial_markets/who_cares_who_wins.pdf, accessed 19 March 2011).

US Social Investment Forum, *Report on Socially Responsible Investing Trends in the United States 2007* (Washington, DC: Social Investment Forum, 2008, www. socialinvest.org/about/contact.cfm, accessed 15 February 2011).

Wood, D., and B. Hoff, *Handbook on Responsible Investment Across Asset Classes: Institute for Responsible Investment* (Cambridge, MA: Initiative for Responsible Investment, 2007, hausercenter.org/iri/?page_id=6, accessed 15 February 2011).

World Economic Forum and AccountAbility, *Mainstreaming Responsible Investment* (London: AccountAbility, 2005, www.accountability.org/about-us/ publications/mainstreaming.html, accessed 22 February 2011).

Index

NOTE: Page numbers in *italic figures* refer to tables

About the authors

Céline Louche is Assistant Professor at Vlerick Leuven Gent Management School, Belgium, where she teaches and researches in corporate social responsibility (CSR). She previously worked as Sustainability Analyst for responsible investment at the Dutch Sustainability Research institute (now Jantzi-Sustainalytics). In her work, she explores the way processes of change take place. A major research interest is the construction of the CSR field with a special focus on responsible investment and stakeholder processes. Céline is the editor of *Innovative CSR* (with Samuel O. Idowu and Walter Leal Filho; Greenleaf Publishing, 2009), *Theory and Practice of Corporate Social Responsibility* (with Samuel O. Idowu; Springer, 2011) and Finance and Sustainability (with William Sun and Roland Perez; Emerald, forthcoming). She has a PhD in management and environmental sciences from the Erasmus University Rotterdam in the Netherlands for which she was awarded the 2005 FIR Finance & Sustainability Award. She is member of the academic and management board of the European Academy of Business in Society, the Scientific Committee of the International Network for Research on Organisations and Sustainable Development, and the Register Commission of Forum ETHIBEL.

Steve Lydenberg has worked for over 35 years in the responsible investment field. He has worked with responsible investment money management firms as Chief Investment Officer for Domini Social Investments and Research Associate with Trillium Asset Management and with corporate social responsibility research firms as Research Director for KLD Research & Analytics and the Council on Economic Priorities. He is also the founding director of the Initiative of Responsible Investment at the Harvard Kennedy School.

Steve is author of the books *Corporations and the Public Interest: Guiding the Invisible Hand* (Berrett-Koehler, 2005) and co-author of *Investing for Good* (Harper Collins, 1993) with Peter Kinder and Amy Domini. His articles on responsible investment and corporate social responsible have appeared in the *Journal of Corporate Citizenship, Corporate Governance: An International Review* and the *Journal of Investing*.